Introduction to Mobile Communications and Networks

Introduction to Mobile Communications and Networks

Charles Harper

NY RESEARCH PRESS

New York

Published by NY Research Press
118-35 Queens Blvd., Suite 400,
Forest Hills, NY 11375, USA
www.nyresearchpress.com

Introduction to Mobile Communications and Networks
Charles Harper

International Standard Book Number: 978-1-63238-862-9 (Hardback)

Cataloging-in-Publication Data

Introduction to mobile communications and networks / Charles Harper.
    p. cm.
Includes bibliographical references and index.
ISBN 978-1-63238-862-9
1. Mobile communication systems. 2. Wireless communication systems.
3. Communication--Network analysis. I. Harper, Charles.
TK5103.2 .I58 2022
621.384--dc23

TABLE OF CONTENTS

Preface   VII

Chapter 1   Introduction   1

- Mobile Communication   1
- Mobile Network   5
- Call Admission Control in Cellular Networks   8

Chapter 2   Mobile Communication Standards   33

- GSM   33
- CDMA   35
- GPRS   43
- FDMA   49
- SDMA   52
- LTE   53
- Voice Over LTE   62

Chapter 3   Cellular Network Technology   64

- Zero Generation   64
- First Generation   65
- Second Generation   66
- Third Generation   69
- Fourth Generation   88
- Fifth Generation   97

Chapter 4   Mobile Network Architecture   100

- Cell Tower   100
- Base Transceiver Station   112
- Cell Phone Signal Booster   115
- Base Station Subsystem   122
- Cellular Repeater   126

- GSM Network Architecture   128
- GPRS Architecture   139
- LTE Network Architecture   141

Chapter 5	Mobile Network: Threats, Attacks and Security	171

- Mobile Network Threats   171
- Signalling Attacks in Mobile Networks   175
- Attacks on LTE in 4G   181
- IP Spoofing in Mobile Network   183
- Mobile Device Security   190
- Security Threat and Countermeasure on 3G Network   191
- Security of Software Defined Mobile Networks   196

Permissions

Index

PREFACE

This book has been written, keeping in view that students want more practical information. Thus, my aim has been to make it as comprehensive as possible for the readers. I would like to extend my thanks to my family and co-workers for their knowledge, support and encouragement all along.

Mobile network is a communication network which is distributed over land areas by cells. These cells are provided with network coverage by base transceiver stations. These can be used further for transmission of data and voice. Mobile communication helps us in communicating with others in different locations without the use of wires and cables. Mobile networks are the backbone of telecommunications. There are several standards of digital cellular networks. These include GSM, CDMA, VoLTE, etc. The set of frequency ranges which fall within the ultra high frequency band are known as cellular frequencies. They are assigned for mobile phones so that it can connect it to cellular networks. These networks have evolved through generations. These generations are 1G, 2G, 3G and 4G, and the latest technology is 5G. This book elucidates the concepts and innovative models around prospective developments with respect to mobile communications and networks. It explores all the important aspects of this field in the present day scenario. The book is appropriate for those seeking detailed information in this area.

A brief description of the chapters is provided below for further understanding:

Chapter – Introduction

The technology which allows us to communicate with people in distant places, without using physical wires or cables as the medium, is referred to as mobile communication. A network that uses wireless mode of communication is termed as mobile network. This is an introductory chapter which will briefly introduce all the significant aspects of mobile communication and networks.

Chapter – Mobile Communication Standards

Mobile communication standard defines a set of protocols used by digital mobile cellular networks. Some of these standards are GSM, CDMA, GPRS, FDMA, OFDMA, SDMA, LTE, LTE advanced, voice over LTE, etc. The topics elaborated in this chapter will help in gaining a better perspective about these different mobile communication standards.

Chapter – Cellular Network Technology

There are a number of technologies which are used in cellular network to establish seamless communication. Some of these network technologies include zero generation, first generation, second-generation, third generation, fourth generation and fifth generation technology. This chapter closely examines these technologies of cellular network to provide an extensive understanding of the subject.

Chapter – Mobile Network Architecture

Mobile network architecture consists of different elements which are necessary for mobile communication. Cell tower, base transceiver station, cell phone signal booster, base station subsystem, cellular repeater, etc. are some of the elements that fall under its domain. This chapter delves into different elements of mobile network architecture which will provide an easy understanding of the subject.

Chapter – Mobile Network: Threats, Attacks and Security

Mobile network is prone to different kinds of threats with the increasing number of users and operators. Data leakage, network spoofing, phishing attacks, broken cryptography, etc. are some of the threats to network security. This chapter has been carefully written to provide an in-depth understanding of these threats, attacks and security of mobile networks.

Charles Harper

Introduction 1

- **Mobile Communication**
- **Mobile Network**
- **Call Admission Control in Cellular Networks**

The technology which allows us to communicate with people in distant places, without using physical wires or cables as the medium, is referred to as mobile communication. A network that uses wireless mode of communication is termed as mobile network. This is an introductory chapter which will briefly introduce all the significant aspects of mobile communication and networks.

Mobile Communication

Mobile Communication is the use of technology that allows us to communicate with others in different locations without the use of any physical connection (wires or cables). Mobile communication makes our life easier, and it saves time and effort.

A mobile phone (also called mobile cellular network, cell phone or hand phone) is an example of mobile communication (wireless communication). It is an electric device used for full duplex two way radio telecommunication over a cellular network of base stations known as cell site.

Features of Mobile Communication

The following are the features of mobile communication:

- High capacity load balancing: Each wired or wireless infrastructure must incorporate high capacity load balancing.

  High capacity load balancing means, when one access point is overloaded, the system will actively shift users from one access point to another depending on the capacity which is available.

- Scalability: The growth in popularity of new wireless devices continuously increasing day by day. The wireless networks have the ability to start small if necessary, but expand in terms of coverage and capacity as needed - without having to overhaul or build an entirely new network.

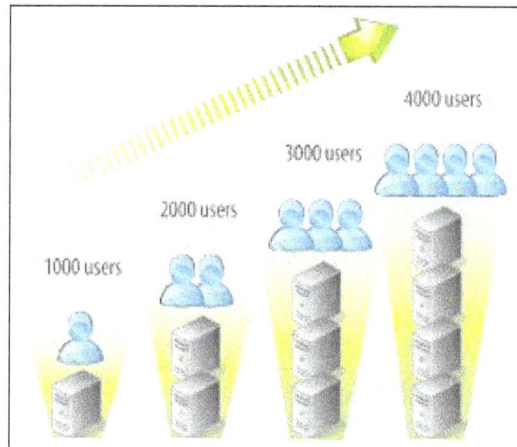

- Network management system: Now a day, wireless networks are much more complex and may consist of hundreds or even thousands of access points, firewalls, switches, managed power and various other components. The wireless networks have a smarter way of managing the entire network from a centralized point.

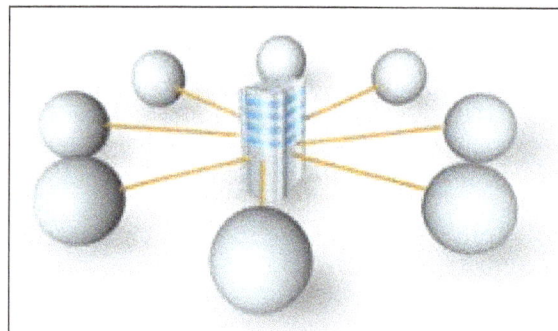

- Role based access control: Role based access control (RBAC) allows you to assign roles based on what, who, where, when and how a user or device is trying to access your network. Once the end user or role of the devices is defined, access control policies or rules can be enforced.

○ Indoor as well as outdoor coverage options: It is important that your wireless system has the capability of adding indoor coverage as well as outdoor coverage.

○ Network access control: Network access control can also be called as mobile device registration. It is essential to have a secure registration. Network access control (NAC) controls the role of the user and enforces policies. NAC can allow your users to register themselves to the network. It is a helpful feature that enhances the user experience.

○ Mobile device management: Suppose, many mobile devices are accessing your wireless network; now think about the thousands of applications are running on those mobile devices. Mobile device management can provide control of how you will manage access to programs and applications. Even you can remotely wipe the device if it is lost or stolen.

• Roaming: You don't need to worry about dropped connections, slower speeds or any disruption in service as you move throughout your office or even from building to building wireless needs to be mobile first. Roaming allows your end-users to successfully move from one access point to another without ever noticing a dip in a performance. For example, allowing a student to check their mail as they walk from one class to the next.

• Redundancy: The level or amount of redundancy your wireless system requires depends on your specific environment and needs. For example: A hospital environment will need a higher level of redundancy than a coffee shop. However, at the end of the day, they both need to have a backup plan in place.

• Proper Security means using the right firewall: The backbone of the system is your network firewall. With the right firewall in place you will be able to:

○ See and control both your applications and end users.

○ Create the right balance between security and performance.

- o   Reduce the complexity with:

    - Antivirus protection.

    - Deep Packet Inspection (DPI).

    - Application filtering.

- o   Protect your network and end users against known and unknown threads including:

    - Zero- day.

    - Encrypted malware.

    - Ransom ware.

    - Malicious botnets.

- Switching: Basically, a network switch is the traffic cop of your wireless network which making sure that everyone and every device gets to where they need to go. Switching is an essential part of every fast, secure wireless network for several reasons:

    - o   It helps the traffic on your network flow more efficiently.

    - o   It minimizes unnecessary traffic.

    - o   It creates a better user experience by ensuring your traffic is going to the right places.

Advantages and Disadvantages of Mobile Communication

Mobile communication technology includes devices such as cellular phones, Wi-Fi-enabled hand-held devices and wireless laptops that can connect through Wi-Fi or with a cellular connection. Consumers envision the benefits of these kinds of devices before purchasing them and sign a provider carrier contract. But it is important to understand the advantages and disadvantages of mobile communication technology before getting involved in a long-term agreement.

Emergencies

A mobile communication device can be helpful in case of an emergency. If emergency authorities are needed, then the cellular phone can be used to contact them. Communications devices with built-in digital phones can be used to catalogue the events at the scene of an accident to help determine responsibility and assess damage. Hikers can take mobile communication devices with them and use the GPS tracking system to find their way, or call for help if needed.

Sharing Information

The Internet has helped broaden communication channels by connecting people all over the world through a single computer network. But before the development of mobile communication devices, the information still needed to be transported back to a computer before it could be sent out over the Internet. With hand-held communications devices, business professionals can instantly share information with clients and vendors regardless of where they are, and friends can share photographs and messages instantly without having to wait until they are logged in to a computer.

Safety Concerns

The use of mobile communication devices can be dangerous. 32 percent of men and 25 percent of women surveyed indicated that they do not drive as safely as they could because of hand-held communication device distractions. Driving a car or safely crossing the street can become difficult when a communication device is causing a distraction.

Less Down Time

Because many business professionals are connected to clients and business associates through cellular devices, there is no down time anymore. Business managers, small business owners and professionals are always on call to clients because of the ability of clients to reach business professionals through cellular phone calls, texting or emails. The same mobile communication tools that can make business easier, can also make business a burden when they take away time off.

Mobile Network

A mobile network is a complex web of connected cellphone tower zones. Mobile networks are also known as cellular networks. They're made up of "cells," which are areas of land that are typically hexagonal, have at least one transceiver cell tower within their area, and use various radio frequencies. These cells connect to one another and to telephone switches or exchanges. Cell towers connect to each other to hand off packets of signals — data, voice, and text messaging — ultimately bringing these signals to mobile devices such as phones and tablets that act as receivers.

Providers use each other's towers in many areas, creating a complex web that offers the widest possible network coverage to subscribers.

Mobile networks have become the backbone of telecommunications, with the widespread adoption of smartphones, tablets, and other mobile devices.

Frequencies

Many network subscribers can use mobile networks' frequencies at the same time. Cell tower sites and mobile devices manipulate the frequencies so that they can use low-power transmitters to supply their services with the least possible interference.

3G, 4G and 5G Networks

Mobile networks have evolved through a series of generations, each representing significant technological improvements over the previous generations. The first two generations of mobile networks first introduced analog voice (1G) and then digital voice (2G). Subsequent generations supported the proliferation of smartphones by introducing data connections (3G) and allowing access to the internet. 4G service networks improved data connections, making them faster and better able to provide greater bandwidth for uses such as streaming.

The latest technology is the 5G network, which promises even faster speeds and greater bandwidth compared with 4G while reducing interference with other nearby wireless devices. Where 4G uses frequencies below 6 GHz, newer 5G networks use shorter wavelength signals with much higher frequencies, in the range of 30 GHz to 300 GHz. These frequencies provide higher bandwidth and allow signals to be more directional, thus reducing interference.

The promise of very high 5G wireless speeds opens the possibility of replacing traditional wired connections to your home, such as cable, with a wireless one, thus greatly expanding the availability of high-speed internet access.

Leading Mobile Network Providers

Cellular service providers in the U.S. range in size from small, regional companies to large, well-known corporations in the telecommunications field, such as Verizon Wireless, AT&T, T-Mobile, US Cellular, and Sprint.

Types of Mobile Networks

The mobile technologies that large mobile service providers use varies, and mobile devices are built to use the technology of the intended carrier and region. The two main mobile technologies in use are Global System for Mobile communications (GSM), which is an international standard, and Code Division Multiple Access (CDMA), owned by Qualcomm. GSM phones don't work on CDMA networks, and vice versa. Long-Term Evolution (LTE) is based on GSM and offers greater network capacity and speed.

GSM vs. CDMA Mobile Networks

Signal reception, call quality, and speed all depend on many factors. The user's location, service provider, and equipment all play a role. GSM and CDMA don't differ much on quality, but the way they work does.

From a consumer standpoint, GSM is more convenient because a GSM phone carries the entire customer's data on a removable SIM card; to change phones, the customer simply swaps the SIM card into the new GSM phone, and it connects to the provider's GSM network. A GSM network must accept any GSM-compliant phone, leaving consumers quite a bit of freedom over their choices in equipment.

CDMA phones, on the other hand, aren't as easily transferred between carriers. CDMA carriers identify subscribers based on whitelists, not SIM cards, and only approved phones are allowed on their networks. Some CDMA phones have SIM cards, but these are for the purpose of connecting to LTE networks or for flexibility when the phone is used outside of the U.S.

GSM wasn't available in the mid-1990s when some networks switched from analog to digital, so they locked into CDMA — at the time, the most advanced mobile network technology.

Advantages and Disadvantages

Benefits or Advantages of Cellular Network

Following are the benefits or advantages of Cellular Network:

- It provides voice/data services even while roaming.

- It connects both fixed and wireless telephone users.

- It is used in areas where cables cannot be laid out due to its wireless nature.

- It is easy to maintain.

- It is easy to upgrade the equipment.

- The mobile and fixed subscribers are connected immediately with cellular network as soon as mobile phones are switched on. All the handshake signals between mobile and base station are automatically exchanged.

Drawbacks or Disadvantages of Cellular Network

Following are the disadvantages of Cellular Network:

- It offers less data rate compare to wired networks such as fiber optics, DSL etc. The data rate varies based on wireless standards such as GSM, CDMA, LTE etc.

- Macro cells are affected by multipath signal loss.

- The capacity is lower and depends on channels/multiple access techniques employed to serve subscribers.

- As the communication is over the air, it has security vulnerabilities.

- It requires higher cost in order to setup cellular network infrastructure.

- The wireless communication is influenced by physical obstructions, climatic conditions and interference from other wireless devices.

- The installation of antennas for cellular network require space and foundation tower. This is very cumbersome and requires both time and effort.

There are different wireless standards used in cellular networks such as GSM, CDMA, LTE etc. Refer following links to know advantages and disadvantages of these technology based cellular networks.

Call Admission Control in Cellular Networks

The service area of a cellular network is divided into cells. Users are connected to base stations in the cells via radio links. Channel frequencies are reused in cells that are sufficiently separated in distance so that mutual interference is below tolerable levels. When a new call is originated in a cell, one of the channels assigned to the base station of the cell is used for communication between the mobile user and the base station (if any channel is available for the call). If all the channels assigned to this base station are in use, the call attempt is assumed to be blocked and cleared from the system (blocked calls cleared). When a new call gets a channel, it keeps the channel until either the call is completed inside the cell or the mobile station (user) moves out of the cell. When the call is completed, the channel is released and becomes available to serve another call.

When a mobile station moves across the cell boundary and enters a new cell, a handover is required. Handover is also named handoff. If an idle channel is available in the destination cell, a channel is assigned to it and the call stays on; otherwise the call is dropped. Two commonly used performance measures for cellular networks are dropping probability of handover calls and blocking probability of new calls. The dropping probability of handover calls represents the probability that a handover call is dropped during handover. The blocking probability of new calls represents the probability that a new call is denied access to the network.

Call admission control (CAC) algorithms are used in order to keep control on dropping probability of handover calls and blocking probability of new calls. They determine whether a call should be accepted or rejected at the base station. Both the blocking probability of new calls and the dropping probability of handover calls are affected by the call admission algorithm used. The call admission algorithms must give priority to handover calls as compared to new calls.

Various priority based call admission algorithms have been reported in the literature, as for example. They can be classified into two basic categories:

- Reservation based schemes: In these schemes, a subset of channels is reserved for exclusive use by handover calls. Whenever the number of calls (new calls) exceeds a certain threshold, these schemes reject new calls until the number of simultaneous calls (new calls) decreases below the threshold. These schemes accept handover calls as long as the cell has idle channels. When the number of calls is compared with the given threshold, this scheme is called call bounding. When the current number of new calls is compared with the given threshold, the scheme is called new call bounding scheme. Equal Access sharing with reservation schemes reserve an integral number of channels or a fractional number channels for exclusive use by handover calls. Schemes with fractional number of guard channels have better control of the blocking probability of the new calls and the dropping probability of the handover calls than schemes with integral number of guard channels.

- Call thinning schemes: These schemes accept new calls with a certain probability that depends on the number of ongoing calls in the cell. New call thinning schemes accept new calls with a probability that depends on the number of ongoing new calls in the cell. Both schemes accept handover calls whenever the cell has free channels.

Dimensioning of Multi-tier Networks

We have many types of multi-tier cellular networks all around us.

Mobile Networks

A multi-tier cellular network is a network that has different types of cells overlaying each other. Each type of cell differs from others by the size. The smaller the size of the cells in a certain area is the more channels are available for users (since the number of channels per cell is fixed). We may consider four types of cells:

- Pico-cells (range 10 – 50 m) are used inside buildings and lifts. The cell antennae are placed in corners of a room or in hallways. Pico-cells are used when the number of users in a building is large and signals from the outside cells cannot penetrate the building. A new type having almost the same features as pico-cells are called femto-cells.

- Micro-cells (range 50 m – 1 km) are cells used mainly in cities where there are a lot of users.

- Macro-cells (range 1 km – 20 km) are used in rural areas since the number of users is small, and in populated areas where micro-cells are too small to handle frequent handovers of users that are moving fast while making calls. For example, if you are in a high speed car and connected to a micro-cell and the car moves too fast for the call to be handed over from one base station (cell antenna) to another, then the call will be dropped.

- Satellites (worldwide coverage).

Having multi-tier cellular networks increases the number of cells, which means that more users are able to use the network without being blocked, and that users in cars or any high speed vehicles are able to talk without worrying about their calls being disconnected.

Hybrid Networks

The proliferation of computer laptops, personal digital assistants (PDA), and mobile phones, coupled with the nearly universal availability of wireless communication services is enabling the goal of ubiquitous wireless communications. Unfortunately, to realize the benefits of omni-present connectivity, users must contend with the challenges of a confusing array of incompatible services, devices, and wireless technologies. Rice University, USA, is developing RENÉ, a system that enables ubiquitous and seamless communication services. The key innovations are a first-of-its-kind multi-tier network interface card, intelligent proxies that enable a new level of graceful adaptation in unmodified applications, and a novel approach to hierarchical and coarse-grained quality of service provisioning. The design of RENÉ requires a coordinated, collaborative effort across traditional layers and across different time scales of the system to maintain uninterrupted user connectivity.

Battlefield Networks

Future battlefield networks will consist of various heterogeneous networking systems and tiers with disparate capabilities and characteristics, ranging from ground ad hoc mobile, sensor networks, and airborne-rich sky networks to satellite networks. It is an enormous challenge to create a suite of novel networking technologies that efficiently integrate these disparate systems.

The key result is the application part with the extension of the Equivalent Random Traffic method for estimation of throughput for networks with traffic splitting and correlated streams. The excellent accuracy (relative error less than 1%) is shown by numerical examples. The ERT-method has been developed for planning of alternate routing in telephone systems by many authors: and others.

Analysis of CAC Strategies in Single Tier Networks: Two Channel Case

We compare four basic CAC schemes by examining new call and handover call blocking probabilities:

- Strategy 1: Dynamic reservation: The cutoff priority scheme is to reserve some channels for handover calls. Whenever a channel is released, it is returned to the common pool of channels.

- Strategy 2: Fractional dynamic reservation: The fractional guard channel scheme (the new call thinning scheme) is to admit a new call with a certain probability which depends on the number of busy channels.

- Strategy 3: Static (fixed) reservation: All channels allocated to a cell are divided into two groups: one to be used by all calls and the other for handover calls only (the rigid division-based CAC scheme).

- Strategy 4: New call bounding scheme: Limitation of the number of simultaneous new calls admitted to the network.

We consider an N-channel cell without waiting positions and two Poisson call flows: handover call flow of intensity A and new call flow of intensity B. The holding times are exponentially distributed with mean value equal one. The exponential distribution simplifies the formulae and the fact that handover calls already has been served for some time before entering the cell considered, does not influence the remaining service time, as the exponential distribution is without memory. Our optimization criteria is the same for all schemes: to get the maximum revenue if each served A-call costs K units (K > 1) and each served B-call costs one unit.

Dynamic Reservation Strategy is better than Fractional Dynamic Reservation Strategy

Let us start from the 2nd strategy: fractional dynamic reservation. This strategy seems to be more general than the dynamic reservation strategy, but we shall prove that such statement is not true. The system is modeled by a three-state Markov process having the following parameters: Number of channels N = 2, A and B = call flow intensities, p_i = the probability of accepting B-calls for service in state i.

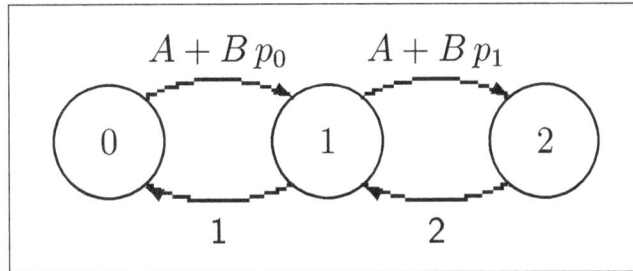

Fractional dynamic reservation: two-channel state transition diagram.

The stationary state probabilities P_i dare defined by equations (up to a normalization factor):

$$P_0 = 1$$
$$P_1 = A + Bp_0$$
$$P_2 = \frac{(A + Bp_0)(A + Bp_1)}{2}$$

The average revenue H equals:

$$H = \frac{A(P_0 + P_1) \cdot K + B(p_0 P_0 + p_1 P_1) \cdot 1}{P_0 + P_1 + P_2}$$
$$= \frac{2(AK + Bp_0 + (A + Bp_0)(AK + Bp_1))}{2 + (A + Bp_0)(2 + A + Bp_1)}.$$

This expression is the ratio of two polynomials, each one with probabilities P_0 and P_1 included in the first power. Consider the above expression as a function of one of the probabilities p. After multiplying by the relevant constant we can get it into the form $(p + a)/(bp + c)$. The derivative of this expression has the form $(ac)/(bp + c)^2$. Consequently, the above expression has invariable sign in the range of values (0, 1), and its extreme values are located at the ends of the interval, i.e. probabilities P_0 and P_1 can only take values 0 or 1. Therefore, the dynamic reservation strategy has advantage over fractional dynamic reservation strategy.

Dynamic Reservation Strategy is better than Static Reservation Strategy

This model is a special case of the previous one: you can reserve 0, 1 or 2 channels, corresponding to choice of probability (p0, p1) in the form of (1,1) (1,0) or (0,0), which, in own order, corresponds to the values of R of 0, 1 or 2. Accordingly, the revenue from formula above takes the form.

$$H_0 = \frac{2(AK+B)(1+A+B)}{2(1+A+B)+(A+B)^2}$$

$$H_1 = \frac{2(B+AK(1+A+B))}{2+(A+B)(2+A)}$$

$$H_2 = \frac{2AK(1+A)}{2+2A+A^2}.$$

How many channels should be reserved? This depends on the parameters K, A and B. To find the optimal value of R, one should solve two equations pointing to the boundary of K:

$$H_0 \quad H_1 \text{ and } H_1 = H_2.$$

We get:

$$K_1 = 1 + \frac{2+A+B}{A(1+A+B)}$$

$$K_2 = 1 + \frac{2(A+1)}{A^2}.$$

It is easy to verify that for any values of A and B, the inequality $K_1 < K_2$ is true since:

$$K_2 - K_1 = \frac{2+2A+2B+A^2+AB}{A^2(1+A+B)}.$$

Hence we have the following solution for optimal reservation R at $N=2$ channels:

$$R = 0 \quad \text{if} \quad K < K_1$$
$$R = 1 \quad \text{if} \quad K_1 < K < K_2$$
$$R = 2 \quad \text{if} \quad K_2 < K$$

Static reservation

In the cases $R=0$ and $R=2$, this strategy does not differ from the strategy 1 above. Therefore, it remains to consider the case $R=1$. This Markov model has four states:

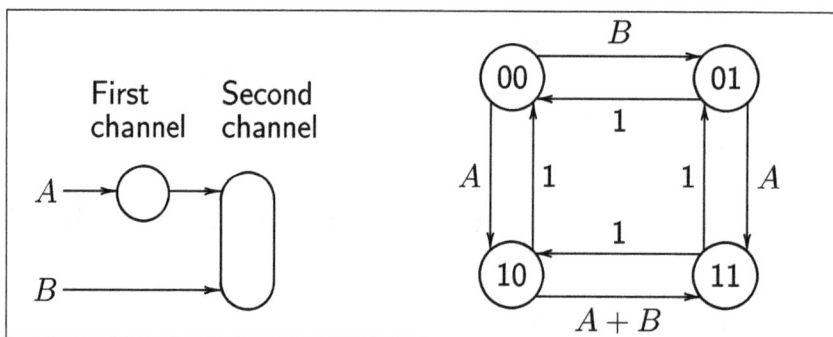

Two channels, static reservation: the model and the state transition diagram.

- (00) – both channels are free,

- (01) – the common channel is engaged,

- (10) – the guard channel is engaged,

- (11) – both channels are engaged.

From linear balance equations under the assumption of statistical equilibrium we obtain the state probabilities (up to a normalizing factor):

$$P_{00} = 2 + 2A + B$$
$$P_{01} = (A + B)^2 + 2B$$
$$P_{10} = A(2 + A + B)$$
$$P_{11} = A((A + B)^2 + A + 2B).$$

The average revenue takes the form:

$$H_4 = \frac{A(2 + 4A + 3B + 2A^2 + 3AB + B^2) \cdot K + B(2 + 4A + B + A^2 + AB) \cdot 1}{2 + 4A + 3B + 3A^2 + 5AB + B^2 + A^3 + 2A^2B + AB^2}$$

Our aim is to prove that the static reservation strategy cannot be more profitable than the dynamic reservation strategy. That is, we must prove that for any values of A, B and K at least one of the values of H_0 and H_1 are not smaller than H_4.

It is easy to verify that by replacing K in this formula by K_1 from expression $K_1 = 1 + \dfrac{2 + A + B}{A(1 + A + B)}$, $K_2 = 1 + \dfrac{2(A+1)}{A^2}$ we obtain $H_4 = 2$. We get the same result by substituting this value of K_1 for K in expressions $H_0 = \dfrac{2(AK + B)(1 + A + B)}{2(1 + A + B) + (A + B)^2}$ and $H_1 = \dfrac{2(B + AK(1 + A + B))}{2 + (A + B)(2 + A)}$, $H_2 = \dfrac{2AK(1 + A)}{2 + 2A + A^2}$,

i.e. the equalities $H_0 = H_1 = H_4 = 2$ are true for any A, B and given $K = K_1$. Furthermore, we note

that H0, H1 and H4 are linear functions of K, i.e. straight lines. Note that for $K < K_1$ the inequalities $H_0 < H_4 < H_1$ are true. Hence these three straight lines intersect at one point, and the straight line H_4 is located between the two others. This means that for any values of A, B and K, at least one of the values of H_0 and H_1 is not less than the value H_4, q.e.d.

New Call Bounding Scheme (Strategy 4)

In case of a 2-channel system, the only nontrivial variant of strategy 4 (Restriction on number of B-calls admitted) is: no more than one B-call. The state transition diagram of this model looks similar to that in Figure, and the expression for average revenue is:

$$H_5 = \frac{2(A(A + B + 1) \cdot K + B(A + 1))}{2(A + B + 1) + A(A + 2B)}$$

This strategy is similar to strategy 3 (Static reservation). As shown in figure, the straight lines H_4 and H_5 are close: up to point $K < 1.25$ strategy 4 is a little more profitable, and from $1.25 < K$ strategy 3 is more profitable. But for any K, strategy 4 is worse than the optimal strategy 1, since $H_5 < H_0$ up to $K < 1.25$ and $H_5 < H_2$ from $1.25 < K$.

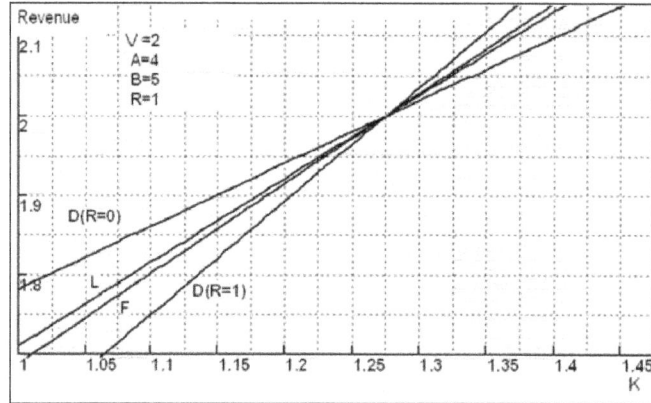

Dependence of the revenue from handover call cost K for three models: D —dynamic reservation, F – static reservation, and L – restriction on number of admitted B-calls.

Hence it has been proved mathematically that the optimal service strategy in a two-channel system is dynamic reservation. Graphically, this fact is illustrated in figure.

- Reservation R = 0, optimal for values of K ≤ 1.25, reservation line D(R = 0),

- Reservation R = 1, optimal for values of K ≥ 1.25, reservation line D(R = 1).

Comparison of Four Strategies in Single Tier Network: Common Case

Fractional Dynamic Reservation

Theorem: For a N-channel loss system where B-calls are accepted with probability P_i, depending on the number of busy channels i (i = 0, 1, . . . , N), the optimal fractional dynamic reservation is limited to probabilities P_i equal to 0 or 1.

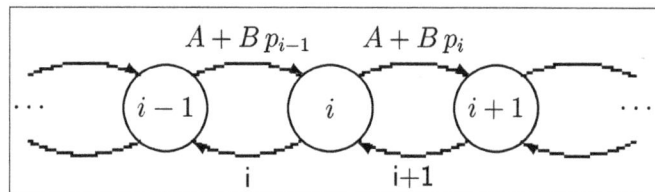

Fractional dynamic reservation: the common case.

The stationary state probabilities F_i are defined by equations (up to normalization factor):

$$F_0 = 1,$$
$$F_1 = A + Bp_0,$$
$$F_2 = (A + Bp_0)(A + Bp_1)/2,$$
$$\text{... ...}$$
$$F_N = (A + Bp_0)(A + Bp_1)...(A + Bp_{N-1})/N!$$

The lost revenue is equal to:

$$C = \frac{B \cdot (F_0 p_0 + F_1 p_1 + \ldots + F_{N-1} p_{N-1}) + (AK + B) \cdot F_N}{F_0 + F_1 + \ldots + F_N}$$

We should maximize the average revenue:

$$H = A \cdot K + B \cdot 1 - C.$$

This expression is the general case of formula $H = \dfrac{A(P_0 + P_1) \cdot K + B(p_0 P_0 + p_1 P_1) \cdot 1}{P_0 + P_1 + P_2} =$

$\dfrac{2(AK + Bp_0 + (A + Bp_0)(AK + Bp_1))}{2 + (A + Bp_0)(2 + A + Bp_1)}$. As above, the expression $H = A \cdot K + B \cdot 1 - C$ is the ratio of

two polynomials, each of which includes the probability pi in first power. Consider the expression $H = A \cdot K + B \cdot 1 - C$ as a function of the probability p_i for any i. By the same argument as above we prove the Theorem.

Dynamic Reservation as Maximum Revenue Strategy

We compare two strategies: dynamic and static reservation. On the basis of numerical results we have shown that with the optimal reservation R the expected revenue is always higher for the model with the dynamic reservation. Naturally, when the handover call cost increases, then the number of reserved channels will increase. For $K = 2$ the optimal dynamic reservation is $R = 2$, and for $K = 4$ it is $R = 4$. The curves, of course. coincide at the ends of the definition interval when R is equal to 0 or N.

Numerical calculations show that strategy 4, restriction of number of B-calls admitted, is similar to strategy 3. For large values of R these strategies are almost identical, but even with the optimal value of R, strategy 4 has only a slight advantage over strategy 3.

The results of numerical analysis confirm that the optimal strategy is dynamic reservation. This statement is strictly proved in the case of a two-channel system.

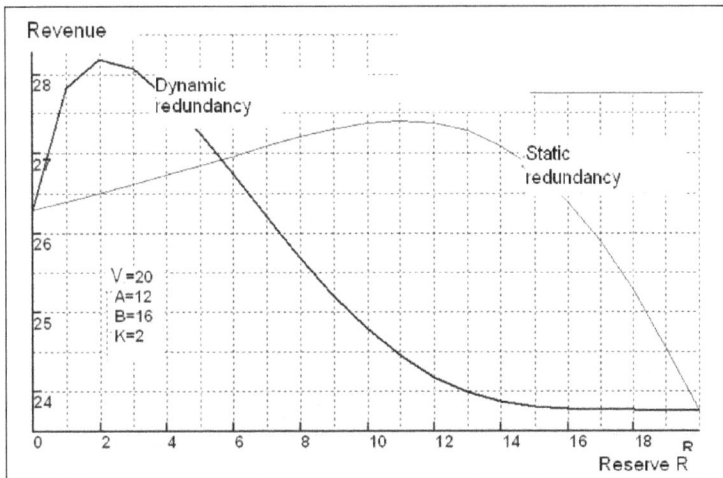

Dependence of revenue on the size of the reservation R for the dynamic reservation D and fixed reservation F. The cost of handover call is $K = 2$.

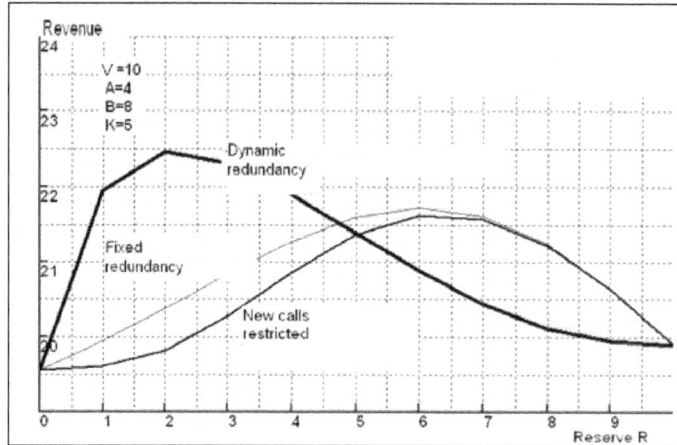

Dependence of revenue on the size of the reservation for three strategies: D = dynamic reservation,
F = static (fixed) reservation, and L = restriction on the number of admitted B-calls.

On Queueing Effect

In queuing priority schemes, new calls and handover calls are all accepted whenever there are idle channels for that type of calls. When no idle channels are accessible, calls may be queued or blocked (i.e cleared from the system). Queueing priority schemes can be divided into three groups: new call queuing schemes, handover call queuing schemes and all calls queuing schemes.

Computational analysis has shown that waiting positions do not change the advantage of dynamic reservation strategy. Figure displays two pairs of curves: one pair is the same as in Figure, the second one relates to a case with three waiting positions. Of course, the revenue is growing, but the preference of dynamic reservation keeps the place.

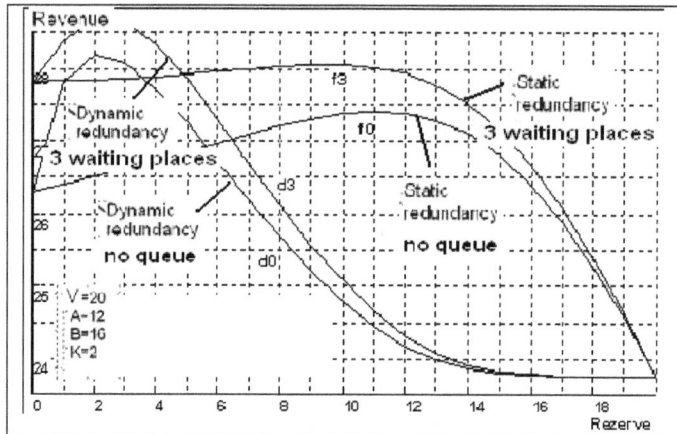

Waiting positions do not change the advantage of dynamic reservation strategy. One pair of curves
is the same as in figure, the other pair relates to the case with three waiting positions.

Two-tier Network

Dynamic Reservation versus Static Reservation

There are several reasons for designing multi-tier cellular networks. One is to provide services for mobile terminals with different mobility and traffic patterns. The required performance measures can be met if the traffic and mobility patterns can be classified into more homogeneous parts and treated separately. Consider a system where there are two mobility classes. If the cell radii are optimized for low-mobility terminals, then the high-mobility terminals will have to make a lot of handovers during a communication session. On the other hand, if the optimization is made regarding the handover performance of high-mobility terminals, then the traffic load in each cell may exceed acceptable limits.

In multi-tier cellular networks, different layers offer the designer the opportunity of class based optimization. In case of a two-tier network we consider three or seven micro-cells overlaid by one large macro-cell offered calls from two mobility classes.

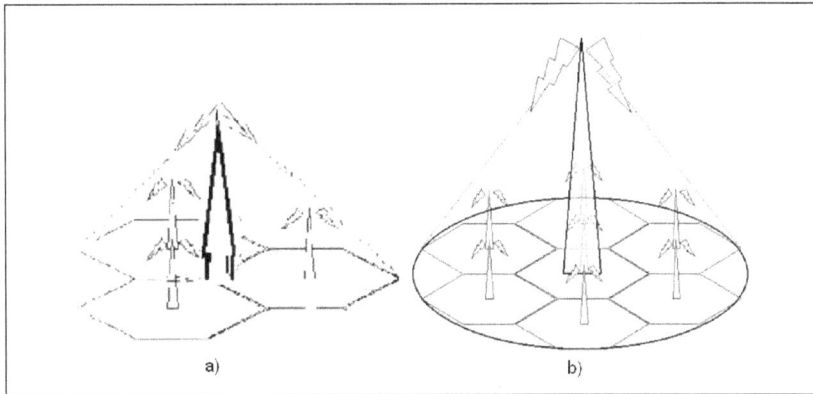

A macro-cell covering (a) three and (b) seven micro-cells.

High-mobility calls of intensity A are served by macro-cell only. Low-mobility calls of intensity B are served by the micro-cells as first choice and, if the reservation strategy admits it, by the macro-cell as second choice. Arriving calls are served as follows. The mobility class of the call is identified. High-mobility calls of intensity A are served by macro-cell only. Low-mobility calls of intensity B are served by the appropriate micro-cell as first choice, and if reservation strategy allows it by macro-cell as second choice. In both cases, our optimization criteria is the same: to maximize the revenue when each served A-call costs K units and each served B-call costs one unit (K > 1).

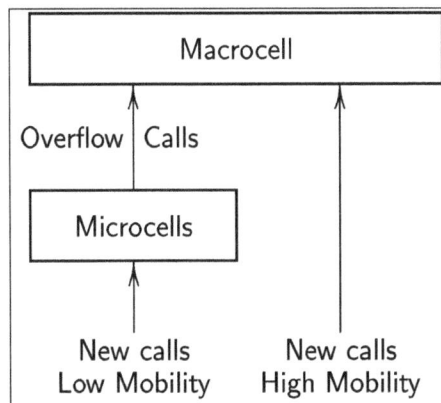

A two-tier cellular network call flow model.

When calls reach the macro-cell level, they are no longer differentiated according to their mobility classes. Therefore, the calls of the high mobility class terminals and the overflowed handover calls from micro-cells are treated identically. New calls from micro-cells may not use the guard (reserved) channels upon their arrival. If no non-guard channel is available, then new calls are blocked. High mobility calls are blocked if all macro-cell channels are busy. Figure shows schematically how calls are served and what order is followed when serving them. As above in the case of single-tier network we compare two reservation strategies:

- Dynamic reservation: The cutoff priority scheme is to reserve a number of channels for high-mobility calls in the macro-cell. Whenever a channel is released, it is returned to the common pool of channels.

- Static reservation: Divide all macro-cell channels allocated to a cell into two groups: one for the common use by all calls and the other for high-mobility calls only (the rigid division-based CAC scheme).

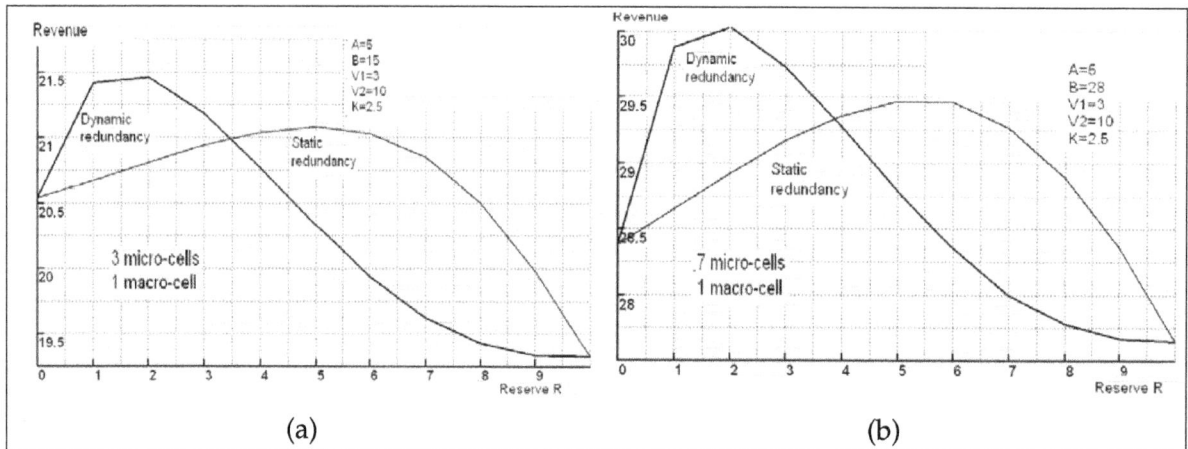

Dependence of the revenue on the reserved number of channels for two-tier networks:
(a) Three micro-cells and one macro-cell, (b) Seven micro-cells and one macro-cell.

Numerical results for two two-tier network examples are obtained. They are not qualitatively different from the results of the one-tier model discussed above. Figure (three micro-cells and one macro-cell) and Figure (seven micro-cells and one macro-cell) show that the dynamic reservation strategy gives the higher maximum revenue in both cases if the reserved number of channels R is properly chosen. The parameters are as follows: A = high-mobility call flow, B = low-mobility call flow, N_1 = number of micro-cell channels for each cell, N_2 = number of micro-cell channels.

In case of two-tier network, the results of numerical analysis confirm that the optimal strategy is dynamic reservation.

Channel Rearrangement Effect

In hierarchical overlaying cellular networks, traffic overflow between the overlaying tiers is used to increase the utilization of the available capacity. The arrival process of overflow traffic has been verified to be correlated and bursty. This characteristic has brought great challenges to performance evaluation of hierarchical networks. In most published works, the discussion is focussed on traffic loss analysis in homogenous hierarchical networks, e.g. micro/macro cellular phone systems as in the numerical analysis below. In the paper, the authors address the problems of performance evaluation in more complicated scenarios by taking account of heterogeneity and user mobility in hierarchical networks. They present an approximate analytical loss model. The loss performance obtained by our approximated analytical model is validated by simulation in a heterogeneous multi-tier overlaying system.

Figure shows the dependence of revenue on channel rearrangement from macro-cell to micro-cell. We are looking for maximum revenue when low-mobility calls cost one unit and high-mobility calls cost K = 3 units.

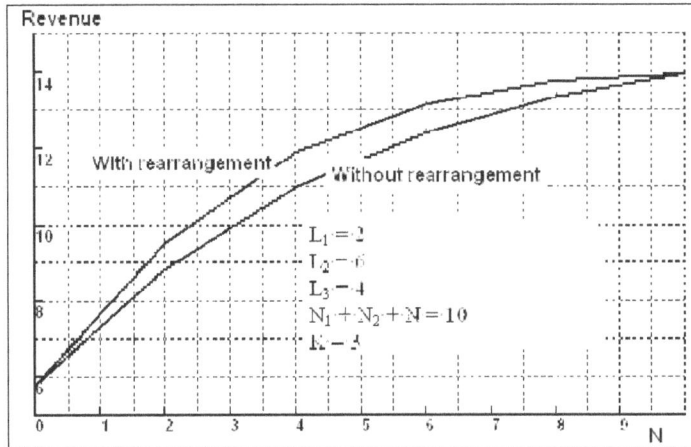

Dependence of revenue on channel rearrangement from macro-cell to micro-cell.

On Optimal Channel Distribution

Figure shows two arrangements each of 18 radio channels for use by 4 call streams. Figure shows a two-tier network with 3 individual channels per stream in the first tier and 6 common channels in the second tier. In Figure some kind of a homogeneous single tier network is depicted: each call has access to 9 channels equally distributed between streams. Such kind of arrangement could be implemented by modern DSP techniques.

Figure depicts the loss probability curves for these two schemes. Case (a) relates to pure loss system, case (b) relates to scheme with one waiting positions per stream. What is surprising? In case (a), beginning with a loss probability as low as 0.56% (less than 1%), it is advantageous to use the equally distributed scheme. Therefore, the traditional two-tier network could be recommended here at a very low call rates. Table 1 contains more detailed data on loss probability. When a single waiting position is added, the advantage of the equally distributed scheme increases even more and the cross point of curves occurs at the loss probability equal to 0.025%.

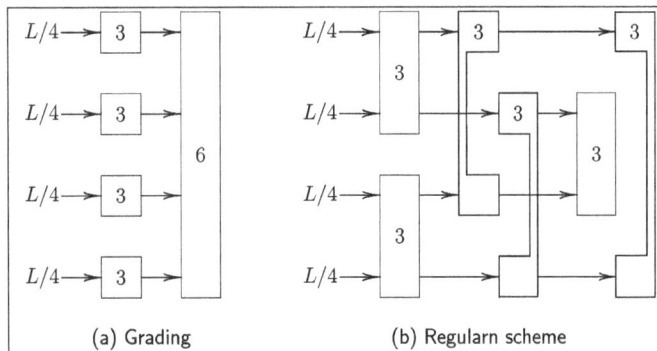

(a) Grading type two-tier network and (b) single-tier network with equally distributed channels.

Table: Loss probabilities for the two schemes of figure (no waiting positions).

Load	Grading	Regular
1	1.0028 10-E11	6.8573 10E-11
3	5.3389 10E-7	1.4922 10E-6
7	1.9967 10E-3	2.1671 10E-3
11	4.0731 10E-2	3.6815 10E-2
15	0.13967	0.12676
20	0.27585	0.25885
25	0.38662	0.37100

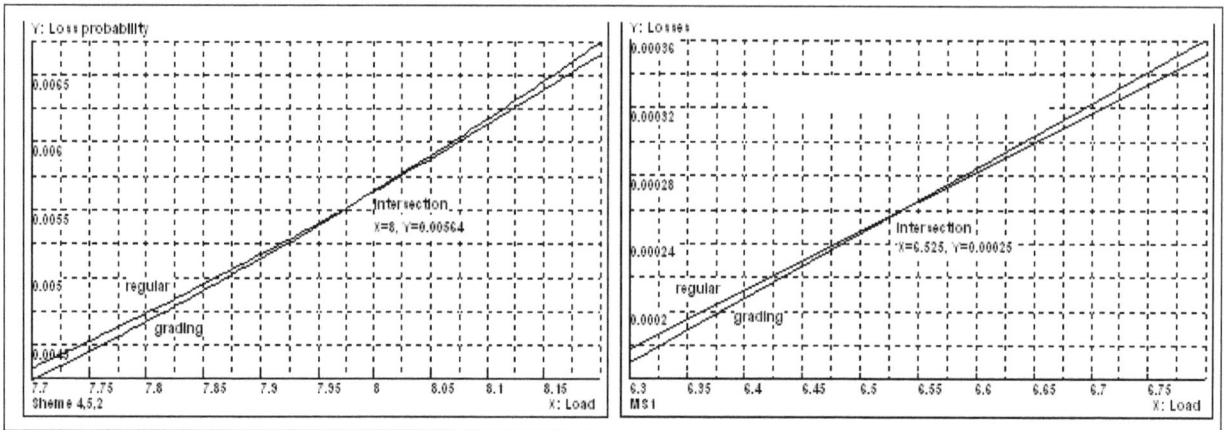

(a) Comparison of channel arrangement between cells for loss system: grading type two-tier network is preferable for low rates only, but as load grows the scheme becomes preferable; (b) the same for the case of one waiting position per stream.

A historical remark regarding this strange phenomenon. It goes almost 50 years back, to the results of V.E. Beneš, a distinguished mathematician from Bell Laboratories, who worked on modeling so-called crossbar telephone switches. In he writes: "The question has arisen, whether there are examples of pairs of networks, with the same number of cross-points, the first of which is better than the second at one value of L, while the second is better than the first at another value of L". The same type of problem was a goal of the studies, but only for the earlier telephone exchange generation with step-by-step switches. In any case, the question remains related to preference of multi-tier networks in comparison with equally distributed schemes, taking modern digital signal processing (DSP) techniques into account.

Multi-tier Network Dimensioning by Equivalent Random Traffic Method

Let us recall some ITU-T documents relating to the dimensioning of circuit groups in traditional telephone networks. These documents deal with dimensioning and service protection methods taking traffic routing methods into account. Recommendations E.520, E.521, E.522 and E.524 deal with the dimensioning of circuit groups with high-usage or final group arrangements. Recommendation E.520 deals with methods for dimensioning of single-path circuit groups based on the use of Erlang's formula.

Recommendations E.521 and E.522 provide methods for the dimensioning of simple alternative routing arrangements as the one shown in Figure, where only first and second-choice routes exist, and where all the traffic overflowing from a circuit group is offered to the same circuit group.

Recommendation E.521 provides methods for dimensioning the final group satisfying GoS (Grade-of-Service) requirements for a given capacity of the high-usage circuit groups. Recommendation E.522 advises on how to dimension high-usage groups to minimize the cost of the whole arrangement. Figure shows a two-tier network. For the dimensioning of such simple alternative routing arrangements the ERT-method is applicable.

Recommendation E.524 provides overflows approximations for non-random traffic inputs which allows for the dimensioning of more complex arrangements, e.g. three-tier network with correlated streams as shown in Figure). The extended ERT-method described below relates to this case and could serve as a basis for a revision of Recommendation E.524.

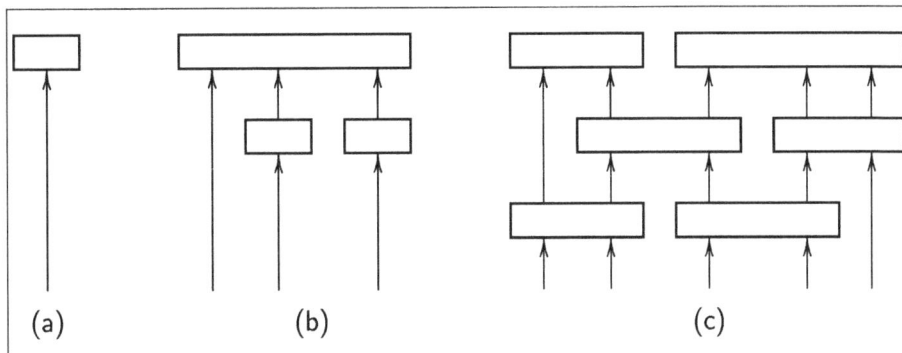

Three network arrangements: single-tier, two-tier, and three-tier with correlated streams.

Erlang Formula and its Generalization for Non-integer Number of Channels

We consider a system of N identical fully accessible channels (servers, trunks, slots, call center agents, pool of wavelengths in the optical network, etc) offered Poisson traffic L and operating as a loss system (blocked calls cleared). The probability that all N channels are busy at a random point of time is equal to:

$$B = E(N, L) = \frac{\dfrac{L^N}{N!}}{\sum_{i=0}^{N} \dfrac{L^i}{i!}}$$

This is the famous Erlang-B formula. For numerical analysis of the above equation we use the well-known recurrent formula:

$$E(N+1, L) = \frac{L \cdot E(N, L)}{N+1+E(N, L)}$$

with initial value $E(0, L) = 1$. For the ERT-method we need Erlang-B formula for non-integral number of channels. How to get solution for a non-integral N? The traditional approach is based on the incomplete gamma function using:

$$E(N, L) = \frac{L^N \cdot e^{-L}}{\Gamma(N+1, L)}$$

Where,

$$\Gamma(N+1,L) = \int_L^\infty t^N e^{-t} dt$$

We propose a new approach for engineering applications. Let the value N be from the interval (0,1). We introduce a parabolic approximation for $R = lnE(N,L)$ at points $N = 0$, $N = 1$ and $N = 2$:

$$\ln E(0,L) = \ln(1) = 0,$$

$$B = \ln E(1,L) = \ln \frac{L}{L+1},$$

$$C = \ln E(2,L) = \ln \frac{L^2}{L^2 + 2L + 2}.$$

Then we get the requested approximation:

$$\ln E(N,L) = \left(\frac{C}{2} - B\right)N^2 + \left(2B - \frac{C}{2}\right)N$$

or in a more convenient form,

$$E(N,L) = L^N (L+1)^{N^2-2N} (L^2 + 2L + 2)^{\frac{N-N^2}{2}}$$

Thus we have an initial value of $E(N,L)$ for N inside the interval (0,1) and we may calculate

$E(N,L)$ at any N by means of recurrent formula $E(N+1,L) = \dfrac{L \cdot E(N,L)}{N+1+E(N,L)}$. In the proposed

approximation $B = E(N,L) = \dfrac{\dfrac{L^N}{N!}}{\sum_{i=0}^N \dfrac{L^i}{i!}}$ is compared numerically with earlier known Erlang-B

formula approximations and it is shown to be more accurate.

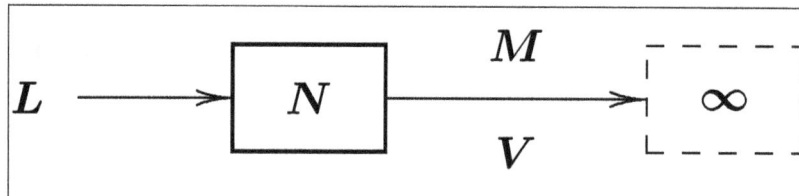

Kosten's model: N fully accessible channels and a channel group of infinite capacity.

Kosten's Model and its New Interpretation

The basic idea the Equivalent Random Traffic (ERT) method is to use Erlang-B formula for overflow traffic offered to a secondary channel group of infinite capacity, the so-called Kosten model. Kosten's paper contains formulae for all binomial moments of number of busy channels in secondary overflow group. In practice, only the two first moments are used for characterization of the overflow traffic: mean traffic intensity M and variance V (reference is usually made to Riordan's paper:

$$M = L \cdot E(N,L)$$

$$V = M \cdot \left(1 - M + \frac{L}{N+1-L+M}\right).$$

From these two parameters one introduce a new parameter Z, the so-called peakedness:

$$Z = \frac{V}{M} = \left(1 - M + \frac{L}{N+1-L+M}\right) \geq 1.$$

Experience shows that peakedness Z is a very good measure for the relative blocking probability a traffic stream with given mean value and variance is subject to.

We offer a new interpretation of Kosten's results. We consider the scheme in figure. There are N common channels and one separate channel. From $M = L \cdot E(N,L)$ and $V = M \cdot \left(1 - M + \frac{L}{N+1-L+M}\right)$ we get a new formula for the variance V when both mean M and mean $M_1 = L \cdot E(N+1,L)$ are known.

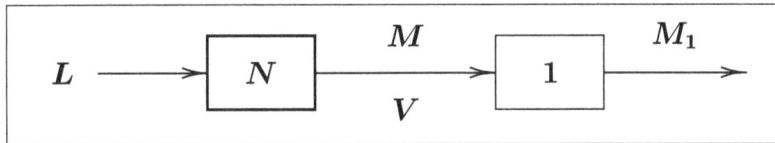

An illustration of the ERT-method extension.

From recurrence formula $E(N+1,L) = \frac{L \cdot E(N,L)}{N+1+E(N,L)}$ follows:

$$M_1 = \frac{LM}{N+1+M}$$

or

$$M + N + 1 = \frac{LM}{M_1}$$

After substitution of $(M+N+1)$ into $V = M \cdot \left(1 - M + \frac{L}{N+1-L+M}\right)$ we get:

$$V = M \cdot \left(1 - M + \frac{L}{\frac{LM}{M_1} - L}\right) = M\left(1 - M + \frac{M}{M - M_1}\right)$$

We can reduce this expression to a simpler form:

$$V = M^2 \left(\frac{1}{M - M_1} - 1 \right) = M \left(\frac{M}{M - M_1} - M \right)$$

Note that $M - M_1$ is the load carried by a single channel and therefore it is always less than one. Formula above is useful for applications of the ERT-method in case of traffic splitting.

ERT-method

The ERT-method has been developed for planning of alternate routing in telephone networks by several authors and others. Figure explains the essence of the method. In g traffic streams which may for example be overflow traffic from other exchanges are offered to a transit exchange. As it is non-Poisson traffic, it cannot be described by classical traffic models. We do not know the distributions (state probabilities) of the traffic streams, but we are satisfied (most often the case in applications of statistics) by describing the i-th traffic stream by its mean value M_i and variance V_i.

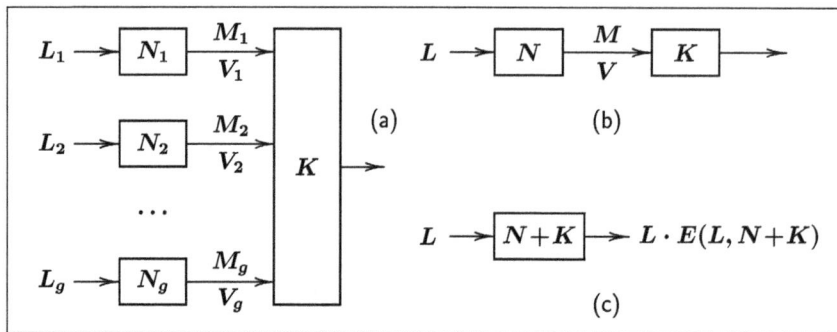

Application of the ERT-method: (a) g independent traffic streams offered to a common group of K channels, (b) equivalent group, (c) Erlang-B formula applied to a common group with $N + K$ channels.

The aggregated overflow process of the g traffic streams is said to be equivalent to the traffic overflowing from a single full accessible group with the same mean and variance as the total overflow traffic. The total traffic offered to the group with K channels has the mean value:

$$M = \sum_{i=1}^{g} M_i$$

We assume that the traffic streams are independent (non-correlated), and thus the variance of the total traffic stream becomes:

$$V = \sum_{i=1}^{g} V_i$$

Therefore, the total traffic is described by M and V. We now consider this traffic to be equivalent to a traffic flow which is lost from a full accessible group and has same mean value M and variance V. For given values of M and V, we therefore solve equations $M = L \cdot E(N,L)$ and

$$V = M \cdot \left(1 - M + \frac{L}{N+1-L+M}\right)$$ with respect to N and L. Then it is replaced by the equivalent

system which is a full accessible system with $(N+K)$ channels offered the traffic L.

On Accuracy of the ERT-method

Let us give a computational analysis of the classical ERT-method by a three-tier network shown in Figure There are four streams each offering a traffic equal to 5 erlang traffic. On first tier there are two servers per stream, on second tier there are three servers, and on third tier two servers. Application of the ERT-method to dimension the alternate routing networks consists of three steps.

- First step is a direct application of formulae $M = L \cdot E(N,L)$ and $V = M \cdot \left(1 - M + \frac{L}{N+1-L+M}\right)$ for two streams and two individual lines (parameters: L = 5, N = 2).

- The second step, we apply the formulae $M = L \cdot E(N,L)$ and $V = M \cdot \left(1 - M + \frac{L}{N+1-L+M}\right)$ to a three-channel group. We get the equivalent parameters L = 9.265, N = 5.83071 and the lost traffic M = 4.3224. The exact value given in brackets is obtained by solving the system of equations of the Markov process, and is equal to 4.30349, i.e. the relative error is less than 1%.

- Third step: The two overflow streams are fed into the two lines. We get the equivalent parameters: L = 16.2076, N = 10.3188, and the lost traffic M = 6.97707. The exact value (in brackets) is 6.91011, i.e. the relative error again is less than 1%.

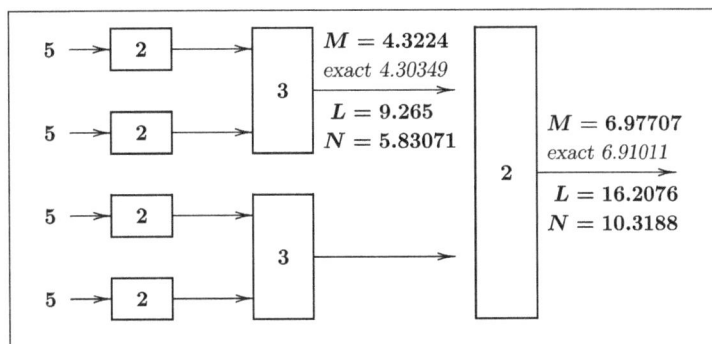

Three-tier network.

The results of calculations show the excellent accuracy of the method. However, such accuracy is not preserved when the number of channels in the third step increases. If instead of two channels we have $(2+g)$ channels in the common group, then Table shows values of the loss for different g values. It is obvious the accuracy drops when increasing g. For a value of g = 10, the relative error is bigger than 3%, but always on the safe side, and the absolute loss probability is very small.

Table: On accuracy of the classical ERT-method.

g	Loss probability Exact	Loss probability ERT-method	Relative error %
0	0.4083	0.4105	0.539
2	0.3239	0.3277	1.173
4	0.2459	0.2506	1.911
6	0.1765	0.1812	2.663
8	0.1181	0.1218	3.133
10	0.0724	0.0747	3.177

Fredericks and Hayward's ERT-method

In an equivalence method is proposed which is simpler to use than Wilkinson-Bretschneider's method. The motivation for the method was first put forward by W.S. Hayward. For given values of (M, V) of a non-Poisson flow, Frederick & Hayward's approach implies direct use of Erlang's formula E(N, M), but with scaling of its parameters as E(N/Z, M/Z). The scaling parameter Z = V/M is the peakedness.

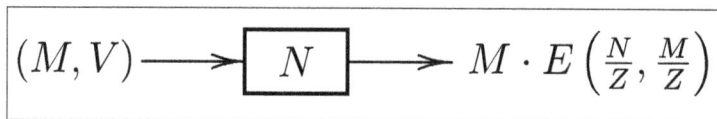

$$(M, V) \longrightarrow \boxed{N} \longrightarrow M \cdot E\left(\frac{N}{Z}, \frac{M}{Z}\right)$$

An illustration of Fredericks & Hayward's approach.

The accuracy of Fredericks & Hayward approach is numerically compared with ERT and with exact values. The calculations were performed for different variants of the scheme shown in Figure. In general, its accuracy is comparable to that of the Wilkinson approach. However, in our opinion Wilkinson's approach is more reliable and always yields worst-case values.

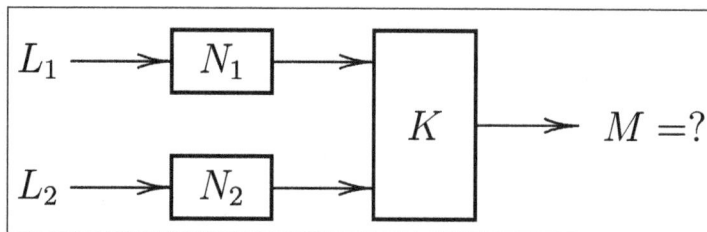

$$L_1 \longrightarrow \boxed{N_1} \longrightarrow$$
$$\boxed{K} \longrightarrow M =?$$
$$L_2 \longrightarrow \boxed{N_2} \longrightarrow$$

On accuracy of Fredericks and Hayward's approach.

Table: On accuracy of Fredericks and Hayward's approach.

L_1	N_1	L_2	N_2	K	M (Wilkinson)	M (Hayward)	M (exact)
1	2	2.5	0	1	1.9740	1.9723	1.9721
6	2	2.5	0	1	5.9635	5.9620	5.9589
8	3	4	4	5	2.8717	2.8514	2.8498
3	3	4	4	5	0.2614	0.2268	0.2608
2	3	4	4	5	0.1122	0.0859	0.1140
4	3	2	4	5	0.1636	0.1386	0.1640
2.5	7	8.25	4	9	0.30268	0.2884	0.30273

Correlation of Overflow Streams

As it is, the ERT-method is not applicable to the analysis of multi-tier networks with correlated streams as shown in figure. In 1960's, this type of problem appeared when dimensioning so-called gradings, the basic structural block in step-by-step exchanges. An important result was developed independently. They determined mean M and variance V of the overflow stream components when split up after a first choice group as shown in Figure On the basis of Kosten's model with two parameters M and V, they developed a 5-parameter model for the two stream case: mean values M_1 and M_2, variances V_1 and V_2, and covariance Cov.

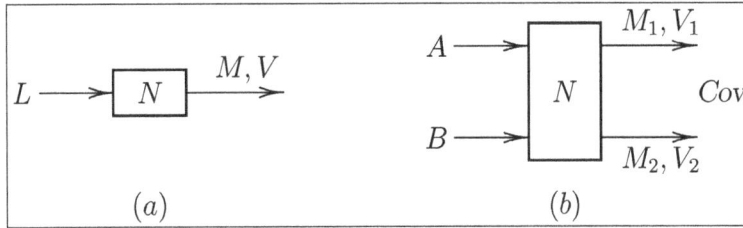

(a) Kosten's model, (b) two-stream model.

If $A = p_1 L$ and $B = p_2 L$, where $p_1 + p_2 = 1$, then the peakedness of partial stream V_i / M_i is defined

by total peakedness V / M :

$$\frac{V_i}{M_i} - 1 = p_i \left(\frac{V}{M} - 1 \right),$$

and covariance

$$Cov = p_1 p_2 (V - M).$$

The covariance formula is proved according to the theorem of variance for mutually dependent variables:

$$V = V_1 + V_2 + 2 \cdot Cov.$$

Correlated Streams: Neal's Formulae

During early 1970's, Scotty Neal from Bell Labs studied the covariance of correlated streams in alternative routing networks. Below we use results from Neal's paper to develop some formulae in notations of figure. We are looking for covariance between two overflow streams after groups with D and F channels, respectively. The key to Neal's solutions is the original work. Neal extended the ERT-method to mutually dependent streams. On basis of the extended Kosten' model he developed a technique for taking correlation into account when combining dependent streams of overflow traffic. More precisely, Neal has considered a 5-parameter Markov model with 5 parameters:

- Number of busy channels in the first choice group (up to N),
- Number of busy channels i in the first alternate group ($0 \le i \le D$),

- Number of busy channels j in the second alternate group ($0 \leq j \leq F$),

- Number of busy channels in the first imaginary infinite channel group,

- Number of busy channels in the second imaginary infinite channel group.

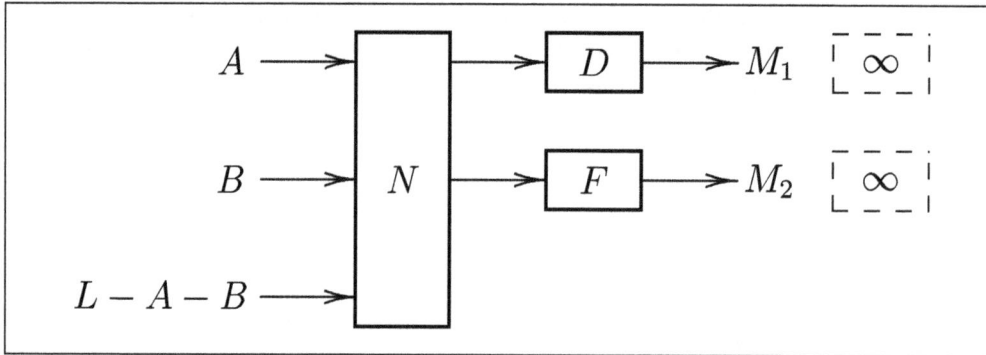

An illustration to Neal's formulas.

After rather sophistical derivations of two-dimensional binomial moment generating functions and using Kosten's approach, Neal obtains linear equations for the two-dimensional probabilities $\beta(i,j), (0 \leq i \leq D, 0 \leq j \leq F)$. The initial value is $\beta(0,0) = E(N,L)$ and other probabilities are defined by linear balance equations:

$$(i + j)v_i + {}_j\beta(i,j) = A\beta(i-1,j) + B \cdot \beta(i,j-1) - A\binom{D}{i-1}\beta(D,j) - B\binom{F}{j-1}\beta(i,F),$$

Where $0 \leq i \leq D, \ 0 \leq j \leq F, \ i+j > 0,$

with the following recurrent formulas for coefficients v_i:

$$v_i = \frac{L}{i \cdot v_{i-1}} + 1 + \frac{N-L}{i}, \quad i > 0, \quad \text{where} \quad v_0 = \frac{1}{E(N,L)}.$$

Then,

$$M_1 = A \cdot \beta(D,0)$$
$$M_2 = B \cdot \beta(0,F)$$
$$Cov = \frac{AQ_1 + BQ_2}{2} - M_1 M_2$$

where,

$$Q_1 = \frac{\sum_{j=0}^{F}\left(\beta(D,j)\prod_{k=j+1}^{F+1}\frac{B}{kv_k}\right)}{1+\sum_{j=1}^{F}\left(\binom{F}{j-1}\prod_{k=j+1}^{F+1}\frac{B}{kv_k}\right)}$$

$$Q_2 = \frac{\displaystyle\sum_{j=0}^{D}\left(\beta(j,F)\prod_{k=j+1}^{D+1}\frac{A}{kv_k}\right)}{1+\displaystyle\sum_{j=1}^{D}\left(\binom{D}{j-1}\prod_{k=j+1}^{D+1}\frac{A}{kv_k}\right)}$$

Comments on Neal's results

In the following discuss the applicability of Neal's results.

Algorithm

We extract the equations which have $j=0$ from the equation system,

$$(i+j)v_i +_j \beta(i,j) = A\beta(i-1,j) + B\cdot\beta(i,j-1) - A\binom{D}{i-1}\beta(D,j) - B\binom{F}{j-1}\beta(i,F),$$

where,

$$0\le i\le D,\ \ 0\le j\le F,\ \ i+j>0.$$

Eliminating members with zero coefficients and considering that $\beta(0,0)$ is known, we obtain the system with D equations referring to $\beta(i,0)$:

$$i\,v_i\beta(i,0) = A\beta(i-1,0) - A\binom{D}{i-1}\beta(D,0).$$

By solving this system of linear equations we get expressions for $\beta(i,0)$:

$$\beta(D,0) = \frac{1}{\displaystyle\sum_{i=0}^{D}\frac{D!}{(D-i)!\,A^i}\prod_{j=0}^{i}v_j},$$

$$\beta(i,0) = \beta(D,0)\binom{D}{i}\left(1+\sum_{j=1}^{D-i}\prod_{k=1}^{j}\frac{v_{i+k}(D+1-i-K)}{A}\right),\ 0<i<D.$$

Values v_i are obtained by formula $v_i = \dfrac{L}{i\cdot v_{i-1}}+1+\dfrac{N-L}{i}$, $\ \ i>0$, where $v_0 = \dfrac{1}{E(N,L)}$. Using direct test we can ascertain that,

$$\beta(D,0) = \frac{1}{\displaystyle\sum_{i=0}^{D}\frac{D!}{(D-i)!\,A^i}\prod_{j=0}^{i}v_j}\ \ \text{and,}$$

$$\beta(i,0) = \beta(D,0)\binom{D}{i}\left(1+\sum_{j=1}^{D-i}\prod_{k=1}^{j}\frac{v_{i+k}(D+1-i-K)}{A}\right),\ 0<i<D,$$

Together with statement $\beta(0,0) = E(N,L)$ indeed satisfies the system of equation,

$$i\,v_i\beta(i,0) = A\beta(i-1,0) - A\binom{D}{i-1}\beta(D,0).$$

The Modified Erlang Formula

Formula $\beta(D,0) = \dfrac{1}{\displaystyle\sum_{i=0}^{D}\dfrac{D!}{(D-i)!\,A^i}\prod_{j=0}^{i}v_j}$ in the form:

$$\beta(D,0) = \dfrac{\dfrac{A^D}{D!}}{\displaystyle\sum_{i=0}^{D}\dfrac{A^{D-i}}{(D-i)!}\prod_{j=0}^{i}v_j}$$

is an obvious analogy to the Erlang formula for the scheme shown in figure. This formula is applicable to the ERT-method. It allows for non-integral number of channels.

From this the mean intensity M_1 follows:

$$M_1 = \dfrac{\dfrac{A^{D+1}}{D!}}{\displaystyle\sum_{i=0}^{D}\dfrac{A^{D-i}}{(D-i)!}\prod_{j=0}^{i}v_j}$$

Formula above is easily implemented and allows for non integer values of N.

Variance

By analogy of equation $V = M\cdot\left(1 - M + \dfrac{L}{\dfrac{LM}{M_1} - L}\right) = M\left(1 - M + \dfrac{M}{M - M_1}\right)$ we get the variance:

$$V_1 = M_1\left(1 - M_1 + \dfrac{M_{1+}}{M_1 - M_{1+}}\right)$$

Where,

$$M_{1+} = A\beta(D+1,0)$$

Substituting A by B and D by F in $M_1 = \dfrac{\dfrac{A^{D+1}}{D!}}{\displaystyle\sum_{i=0}^{D}\dfrac{A^{D-i}}{(D-i)!}\prod_{j=0}^{i}v_j}$ and $V_1 = M_1\left(1 - M_1 + \dfrac{M_{1+}}{M_1 - M_{1+}}\right)$ we get

M2 and V2 in similar way.

Extended ERT-method: Numerical Example

Consider an example where the extended ERT-method can be used and the covariance obtained

by using formulas $\begin{aligned} M_1 &= A \cdot \beta(D,0) \\ M_2 &= B \cdot \beta(0,F) \end{aligned}$ and $Cov = \dfrac{AQ_1 + BQ_2}{2} - M_1 M_2$. Let us calculate the mean

value M of the overflow traffic for the following scheme.

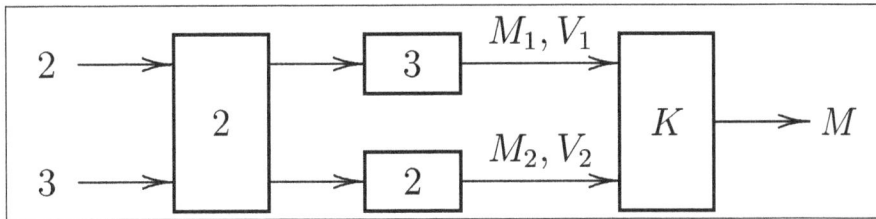

Scheme with correlated streams.

Using formulas $M_1 = \dfrac{\dfrac{A^{D+1}}{D!}}{\displaystyle\sum_{i=0}^{D} \dfrac{A^{D-i}}{(D-i)!} \prod_{j=0}^{i} v_j}$ and $V_1 = M_1\left(1 - M_1 + \dfrac{M_{1+}}{M_1 - M_{1+}}\right)$ we find:

$M_1 = 0.180018$, $M_2 = 0.873221$, $V_1 = 0.252930$, $V_2 = 1.25284$.

Using $\begin{aligned} M_1 &= A \cdot \beta(D,0) \\ M_2 &= B \cdot \beta(0,F) \end{aligned}$ and $Cov = \dfrac{AQ_1 + BQ_2}{2} - M_1 M_2$ we calculate the covariance of the two

streams:

Cov = 0.0282596.

We now can calculate the intensity of the flow which is overflowing to the group with K channels:

$$M^* = M_1 + M_2 = 1.053239, \quad V^* = V_1 + V_2 + 2Cov = 1.56229.$$

Using the extended ERT-method we get the equivalent group:

$$L^* = 3.33306, \quad N^* = 3.44900.$$

Therefore, using Erlang-B formula $M = L^* \cdot E\left(N^* + K, L^*\right)$.

We can obtain mean intensity M of overflow stream for various values of K as shown in Table. The results of calculations show the excellent accuracy of the extended ERT-method.

Table: Accuracy of the Extended ERT-method for correlated streams.

K	M exact	M by ERT	Rel. error%
1	0.63322	0.63801	0.756
2	0.34583	0.34936	1.019

3	0.17067	0.17128	0.357
4	0.07616	0.07492	1.634
5	0.03091	0.02929	5.215

However, such accuracy is not preserved when the number of channels K in the final group increases. Table shows values of the loss for different K values. It is obvious that for increasing K the accuracy drops. For the value $K = 5$ the relative error is greater than 5%. The same effect one observes in Tables. For decreasing (very small) blocking probabilities the accuracy increases, but the absolute error decreases.

References

- Mobile-communication-introduction: javatpoint.com, Retrieved 31 March, 2019

- Advantages-disadvantages-of-mobile-communication-technology: techwalla.com, Retrieved 14 July, 2019

- How-does-a-mobile-network-work: lifewire.com, Retrieved 17 May, 2019

- Advantages-and-disadvantages-of-cellular-network, terminology: rfwireless-world.com, Retrieved 19 April, 2019

Mobile Communication Standards

2

- **GSM**
- **CDMA**
- **GPRS**
- **FDMA**
- **SDMA**
- **LTE**
- **Voice Over LTE**

Mobile communication standard defines a set of protocols used by digital mobile cellular networks. Some of these standards are GSM, CDMA, GPRS, FDMA, OFDMA, SDMA, LTE, LTE advanced, voice over LTE, etc. The topics elaborated in this chapter will help in gaining a better perspective about these different mobile communication standards.

GSM

GSM stands for Global System for Mobile communication. Today, GSM is used by more than 800 million end users spread across 190 countries which represent around 70 percent of today's digital wireless market.

In GSM, geographical area is divided into hexagonal cells whose side depends upon power of transmitter and load on transmitter (number of end user). At the center of cell, there is a base station consisting of a transceiver (combination of transmitter and receiver) and an antenna.

Architecture

Function of Components

- Mobile station (MS): It refers for mobile station. Simply, it means a mobile phone.
- Base Transceiver System (BTS): It maintains the radio component with MS.

- Base station controller (BSC): Its function is to allocate necessary time slots between the BTS and MSC.

- Home location register (HLR): It is the reference database for subscriber parameter ike subscriber's ID, location, authentication key etc.

- Visitor location register (VLR): It contains copy of most of the data stored in HLR which is temporary and exists only until subscriber is active.

- Equipment identity register (EIR): It is a database which contains a list of valid mobile equipment on the network.

- Authentication Center (AuC): It perform authentication of subscriber.

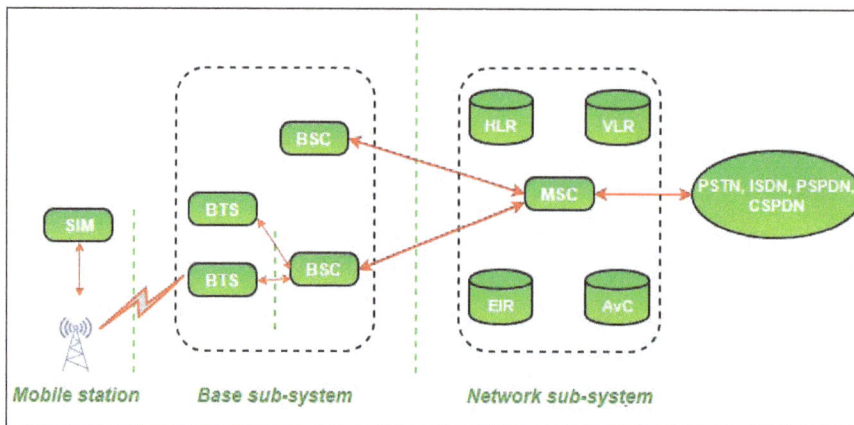

Working

GSM is combination of TDMA (Time Division Multiple Access), FDMA (Frequency Division Multiple Access) and Frequency hopping. Initially, GSM use two frequency bands of 25 MHz width : 890 to 915 MHz frequency band for up-link and 935 to 960 MHz frequency for down-link. Later on, two 75 MHz band were added. 1710 to 1785 MHz for up-link and 1805 to 1880 MHz for down-link. Up-link is the link from ground station to a satellite and down-link is the link from a satellite down to one or more ground stations or receivers. GSM divides the 25 MHz band into 124 channels each having 200 KHz width and remaining 200 KHz is left unused as a guard band to avoid interference.

Control Channels

These are main control channels in GSM:

- BCH (Broadcast Channel): It is for down-link only. It has following types:

  - BCCH (Broadcast Control Channel): It broadcasts information about the serving cell.

  - SCH (Synchronization channel): Carries information like frame number and BSIC (Base Station Identity Code) for frame synchronization.

  - FCCH (Frequency Correction Channel): Enable MS to synchronize to frequency.

- CCCH (Common Control Channel): It has following types:

  ○ RACH (Random Access Channel): Used by MS when making its first access to network. It is for up-link only.

  ○ AGCH (Access Grant Channel): Used for acknowledgement of the access attempt sent on RACH. It is for down-link only.

  ○ PCH (Paging Channel): Network pages the MS, if there is an incoming call or a short message. It is for down-link only.

- DCCH (Dedicated Control Channel): It is for both up-link and down-link. It has following types:

  ○ SDCCH (Stand-alone Dedicated Control Channel): It is used for call setup, authentication, ciphering location update and SMS.

  ○ SACCH (Slow Associated Control Channel): Used to transfer signal while MS have ongoing conversation on topic or while SDCCH is being used.

  ○ FACCH (Fast Associated Control Channel): It is used to send fast message like hand over message.

CDMA

Code-division multiple access (CDMA) is a channel access method used by various radio communication technologies.

CDMA is an example of multiple access, where several transmitters can send information simultaneously over a single communication channel. This allows several users to share a band of frequencies. To permit this without undue interference between the users, CDMA employs spread spectrum technology and a special coding scheme (where each transmitter is assigned a code).

CDMA is used as the access method in many mobile phone standards. IS-95, also called "cdmaOne", and its 3G evolution CDMA2000, are often simply referred to as "CDMA", but UMTS, the 3G standard used by GSM carriers, also uses "wideband CDMA", or W-CDMA, as well as TD-CDMA and TD-SCDMA, as its radio technologies.

The technology of code-division multiple access channels have long been known. In the Soviet Union (USSR), the first work devoted to this subject was published in 1935 by Dmitry Ageev. It was shown that through the use of linear methods, there are three types of signal separation: frequency, time and compensatory. The technology of CDMA was used in 1957, when the young military radio engineer Leonid Kupriyanovich in Moscow made an experimental model of a wearable automatic mobile phone, called LK-1 by him, with a base station. LK-1 has a weight of 3 kg, 20–30 km operating distance, and 20–30 hours of battery life. The base station, as described by the author, could serve several customers. In 1958, Kupriyanovich made the new experimental "pocket" model of mobile phone. This phone weighed 0.5 kg. To serve more customers, Kupriyanovich proposed the device, which he called "correlator." In 1958, the USSR also started the development of the "Altai" national civil mobile phone service for cars, based on the Soviet MRT-1327 standard. The phone system weighed 11 kg (24 lb). It was placed in the trunk of the vehicles of high-ranking officials and used a standard handset in the passenger compartment. The main developers of the Altai system were VNIIS (Voronezh Science Research Institute of Communications) and GSPI (State Specialized Project Institute). In 1963 this service started in Moscow, and in 1970 Altai service was used in 30 USSR cities.

Uses

A CDMA2000 mobile phone:

- One of the early applications for code-division multiplexing is in the Global Positioning System (GPS). This predates and is distinct from its use in mobile phones.

- The Qualcomm standard IS-95, marketed as cdmaOne.

- The Qualcomm standard IS-2000, known as CDMA2000, is used by several mobile phone companies, including the Globalstar network.

- The UMTS 3G mobile phone standard, which uses W-CDMA.

- CDMA has been used in the OmniTRACS satellite system for transportation logistics.

Steps in CDMA Modulation

CDMA is a spread-spectrum multiple-access technique. A spread-spectrum technique spreads the bandwidth of the data uniformly for the same transmitted power. A spreading code is a pseudo-random code that has a narrow ambiguity function, unlike other narrow pulse codes. In CDMA a locally generated code runs at a much higher rate than the data to be transmitted. Data for transmission is combined by bitwise XOR (exclusive OR) with the faster code. The figure shows how a spread-spectrum signal is generated. The data signal with pulse duration of T_b (symbol period) is XORed with the code signal with pulse duration of T_c (chip period). (Bandwidth is proportional to $1/T$, where T = bit time.) Therefore, the bandwidth of the data signal is $1/T_b$ and the bandwidth of the spread spectrum signal is $1/T_c$. Since T_c is much smaller than T_b, the bandwidth of the spread-spectrum signal is much larger than the bandwidth of the original signal. The ratio T_b/T_c is called the spreading factor or processing gain and determines to a certain extent the upper limit of the total number of users supported simultaneously by a base station.

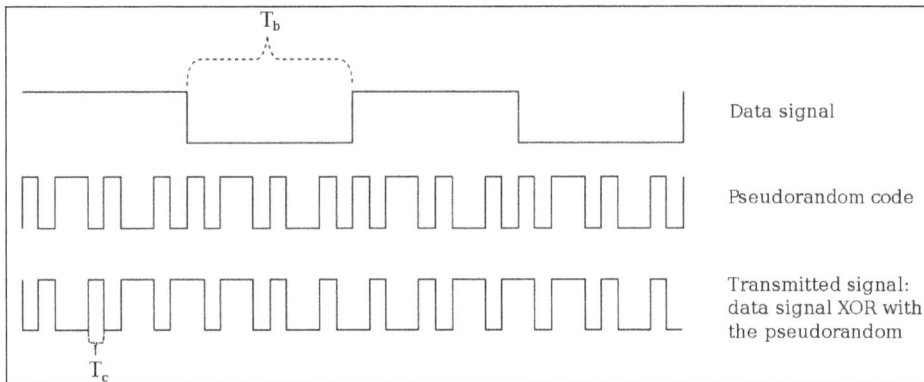

Generation of a CDMA signal.

Each user in a CDMA system uses a different code to modulate their signal. Choosing the codes used to modulate the signal is very important in the performance of CDMA systems. The best performance occurs when there is good separation between the signal of a desired user and the signals of other users. The separation of the signals is made by correlating the received signal with the locally generated code of the desired user. If the signal matches the desired user's code, then the correlation function will be high and the system can extract that signal. If the desired user's code has nothing in common with the signal, the correlation should be as close to zero as possible (thus eliminating the signal); this is referred to as cross-correlation. If the code is correlated with the signal at any time offset other than zero, the correlation should be as close to zero as possible. This is referred to as auto-correlation and is used to reject multi-path interference.

An analogy to the problem of multiple access is a room (channel) in which people wish to talk to each other simultaneously. To avoid confusion, people could take turns speaking (time division), speak at different pitches (frequency division), or speak in different languages (code division). CDMA is analogous to the last example where people speaking the same language can understand each other, but other languages are perceived as noise and rejected. Similarly, in radio CDMA, each group of users is given a shared code. Many codes occupy the same channel, but only users associated with a particular code can communicate.

In general, CDMA belongs to two basic categories: synchronous (orthogonal codes) and asynchronous (pseudorandom codes).

Code-division Multiplexing (Synchronous CDMA)

The digital modulation method is analogous to those used in simple radio transceivers. In the analog case, a low-frequency data signal is time-multiplied with a high-frequency pure sine-wave carrier and transmitted. This is effectively a frequency convolution (Wiener–Khinchin theorem) of the two signals, resulting in a carrier with narrow sidebands. In the digital case, the sinusoidal carrier is replaced by Walsh functions. These are binary square waves that form a complete orthonormal set. The data signal is also binary and the time multiplication is achieved with a simple XOR function. This is usually a Gilbert cell mixer in the circuitry.

Synchronous CDMA exploits mathematical properties of orthogonality between vectors representing the data strings. For example, binary string 1011 is represented by the vector (1, 0, 1, 1). Vectors can be multiplied by taking their dot product, by summing the products of their respective components (for example, if u = (a, b) and v = (c, d), then their dot product u·v = ac + bd). If the dot product is zero, the two vectors are said to be orthogonal to each other. Some properties of the dot product aid understanding of how W-CDMA works. If vectors a and b are orthogonal, then $a \cdot b = 0$ and:

$$a \cdot (a+b) = \|a\|^2, \text{ since } a \cdot a + a \cdot b = \|a\|^2 + 0,$$

$$a \cdot (-a+b) = -\backslash a \backslash^2, \text{ since } -a \cdot a + a \cdot b = \backslash a \backslash^2 + 0,$$

$$b \cdot (a+b) = \|b\|^2, \text{ since } b \cdot a + b \cdot b = 0 + \|b\|^2,$$

$$b \cdot (a-b) = -\|b\|^2, \text{ since } b \cdot a - b \cdot b = 0 - \|b\|^2.$$

Each user in synchronous CDMA uses a code orthogonal to the others' codes to modulate their signal. An example of 4 mutually orthogonal digital signals is shown in the figure below. Orthogonal codes have a cross-correlation equal to zero; in other words, they do not interfere with each other. In the case of IS-95, 64-bit Walsh codes are used to encode the signal to separate different users. Since each of the 64 Walsh codes is orthogonal to all other, the signals are channelized into 64 orthogonal signals. The following example demonstrates how each user's signal can be encoded and decoded.

Example:

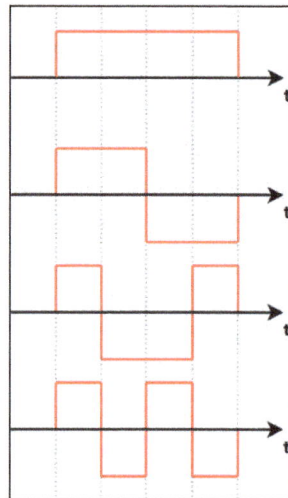

An example of 4 mutually orthogonal digital signals.

Start with a set of vectors that are mutually orthogonal. (Although mutual orthogonality is the only condition, these vectors are usually constructed for ease of decoding, for example columns or rows from Walsh matrices). An example of orthogonal functions is shown in the adjacent picture. These vectors will be assigned to individual users and are called the code, chip code, or chipping code. In the interest of brevity, the rest of this example uses codes v with only two bits.

Each user is associated with a different code, say v. A 1 bit is represented by transmitting a positive code v, and a 0 bit is represented by a negative code −v. For example, if $v = (v_0, v_1) = (1, -1)$ and the data that the user wishes to transmit is (1, 0, 1, 1), then the transmitted symbols would be,

$$(v, -v, v, v) = (v_0, v_1, -v_0, -v_1, v_0, v_1, v_0, v_1) = (1, -1, -1, 1, 1, -1, 1, -1).$$

Each sender has a different, unique vector v chosen from that set, but the construction method of the transmitted vector is identical.

Now, due to physical properties of interference, if two signals at a point are in phase, they add to give twice the amplitude of each signal, but if they are out of phase, they subtract and give a signal that is the difference of the amplitudes. Digitally, this behaviour can be modelled by the addition of the transmission vectors, component by component.

If sender0 has code $(1, -1)$ and data $(1, 0, 1, 1)$, and sender1 has code $(1, 1)$ and data $(0, 0, 1, 1)$, and both senders transmit simultaneously, then this table describes the coding steps:

Step	Encode sender0	Encode sender1
0	code0 = $(1, -1)$, data0 = $(1, 0, 1, 1)$	code1 = $(1, 1)$, data1 = $(0, 0, 1, 1)$
1	encode0 = $2(1, 0, 1, 1) - (1, 1, 1, 1) = (1, -1, 1, 1)$	encode1 = $2(0, 0, 1, 1) - (1, 1, 1, 1) = (-1, -1, 1, 1)$
2	signal0 = encode0 $\otimes$ code0 = $(1, -1, 1, 1) \otimes (1, -1)$ = $(1, -1, -1, 1, 1, -1, 1, -1)$	signal1 = encode1 $\otimes$ code1 = $(-1, -1, 1, 1) \otimes (1, 1)$ = $(-1, -1, -1, -1, 1, 1, 1, 1)$

Because signal0 and signal1 are transmitted at the same time into the air, they add to produce the raw signal:

$$(1, -1, -1, 1, 1, -1, 1, -1) + (-1, -1, -1, -1, 1, 1, 1, 1) = (0, -2, -2, 0, 2, 0, 2, 0).$$

This raw signal is called an interference pattern. The receiver then extracts an intelligible signal for any known sender by combining the sender's code with the interference pattern. The following table explains how this works and shows that the signals do not interfere with one another:

Step	Decode sender0	Decode sender1
0	code0 = $(1, -1)$, signal = $(0, -2, -2, 0, 2, 0, 2, 0)$	code1 = $(1, 1)$, signal = $(0, -2, -2, 0, 2, 0, 2, 0)$
1	decode0 = pattern.vector0	decode1 = pattern.vector1
2	decode0 = $((0, -2), (-2, 0), (2, 0), (2, 0)) \cdot (1, -1)$	decode1 = $((0, -2), (-2, 0), (2, 0), (2, 0)) \cdot (1, 1)$
3	decode0 = $((0 + 2), (-2 + 0), (2 + 0), (2 + 0))$	decode1 = $((0 - 2), (-2 + 0), (2 + 0), (2 + 0))$
4	data0=$(2, -2, 2, 2)$, meaning $(1, 0, 1, 1)$	data1=$(-2, -2, 2, 2)$, meaning $(0, 0, 1, 1)$

Further, after decoding, all values greater than 0 are interpreted as 1, while all values less than zero are interpreted as 0. For example, after decoding, data0 is $(2, -2, 2, 2)$, but the receiver interprets this as $(1, 0, 1, 1)$. Values of exactly 0 means that the sender did not transmit any data, as in the following example:

Assume signal0 = $(1, -1, -1, 1, 1, -1, 1, -1)$ is transmitted alone. The following table shows the decode at the receiver:

Step	Decode sender0	Decode sender1
0	code0 = $(1, -1)$, signal = $(1, -1, -1, 1, 1, -1, 1, -1)$	code1 = $(1, 1)$, signal = $(1, -1, -1, 1, 1, -1, 1, -1)$
1	decode0 = pattern.vector0	decode1 = pattern.vector1
2	decode0 = $((1, -1), (-1, 1), (1, -1), (1, -1)) \cdot (1, -1)$	decode1 = $((1, -1), (-1, 1), (1, -1), (1, -1)) \cdot (1, 1)$
3	decode0 = $((1 + 1), (-1 - 1), (1 + 1), (1 + 1))$	decode1 = $((1 - 1), (-1 + 1), (1 - 1), (1 - 1))$
4	data0 = $(2, -2, 2, 2)$, meaning $(1, 0, 1, 1)$	data1 = $(0, 0, 0, 0)$, meaning no data

When the receiver attempts to decode the signal using sender1's code, the data is all zeros, therefore the cross-correlation is equal to zero and it is clear that sender1 did not transmit any data.

Asynchronous CDMA

When mobile-to-base links cannot be precisely coordinated, particularly due to the mobility of the handsets, a different approach is required. Since it is not mathematically possible to create signature sequences that are both orthogonal for arbitrarily random starting points and which make full use of the code space, unique "pseudo-random" or "pseudo-noise" (PN) sequences are used in *asynchronous* CDMA systems. A PN code is a binary sequence that appears random but can be reproduced in a deterministic manner by intended receivers. These PN codes are used to encode and decode a user's signal in asynchronous CDMA in the same manner as the orthogonal codes in synchronous CDMA. These PN sequences are statistically uncorrelated, and the sum of a large number of PN sequences results in *multiple access interference* (MAI) that is approximated by a Gaussian noise process (following the central limit theorem in statistics). Gold codes are an example of a PN suitable for this purpose, as there is low correlation between the codes. If all of the users are received with the same power level, then the variance (e.g., the noise power) of the MAI increases in direct proportion to the number of users. In other words, unlike synchronous CDMA, the signals of other users will appear as noise to the signal of interest and interfere slightly with the desired signal in proportion to number of users.

All forms of CDMA use spread-spectrum process gain to allow receivers to partially discriminate against unwanted signals. Signals encoded with the specified PN sequence (code) are received, while signals with different codes (or the same code but a different timing offset) appear as wideband noise reduced by the process gain.

Since each user generates MAI, controlling the signal strength is an important issue with CDMA transmitters. A CDM (synchronous CDMA), TDMA, or FDMA receiver can in theory completely reject arbitrarily strong signals using different codes, time slots or frequency channels due to the orthogonality of these systems. This is not true for asynchronous CDMA; rejection of unwanted signals is only partial. If any or all of the unwanted signals are much stronger than the desired signal, they will overwhelm it. This leads to a general requirement in any asynchronous CDMA system to approximately match the various signal power levels as seen at the receiver. In CDMA cellular, the base station uses a fast closed-loop power-control scheme to tightly control each mobile's transmit power.

Advantages of Asynchronous CDMA over Other Techniques

Efficient Practical Utilization of the Fixed Frequency Spectrum

In theory CDMA, TDMA and FDMA have exactly the same spectral efficiency, but, in practice, each has its own challenges – power control in the case of CDMA, timing in the case of TDMA, and frequency generation/filtering in the case of FDMA.

TDMA systems must carefully synchronize the transmission times of all the users to ensure that they are received in the correct time slot and do not cause interference. Since this cannot be perfectly controlled in a mobile environment, each time slot must have a guard time, which reduces the probability that users will interfere, but decreases the spectral efficiency.

Similarly, FDMA systems must use a guard band between adjacent channels, due to the unpredictable Doppler shift of the signal spectrum because of user mobility. The guard bands will reduce the probability that adjacent channels will interfere, but decrease the utilization of the spectrum.

Flexible Allocation of Resources

Asynchronous CDMA offers a key advantage in the flexible allocation of resources i.e. allocation of PN codes to active users. In the case of CDM (synchronous CDMA), TDMA, and FDMA the number of simultaneous orthogonal codes, time slots, and frequency slots respectively are fixed, hence the capacity in terms of the number of simultaneous users is limited. There are a fixed number of orthogonal codes, time slots or frequency bands that can be allocated for CDM, TDMA, and FDMA systems, which remain underutilized due to the bursty nature of telephony and packetized data transmissions. There is no strict limit to the number of users that can be supported in an asynchronous CDMA system, only a practical limit governed by the desired bit error probability since the SIR (signal-to-interference ratio) varies inversely with the number of users. In a bursty traffic environment like mobile telephony, the advantage afforded by asynchronous CDMA is that the performance (bit error rate) is allowed to fluctuate randomly, with an average value determined by the number of users times the percentage of utilization. Suppose there are $2N$ users that only talk half of the time, then $2N$ users can be accommodated with the same *average* bit error probability as N users that talk all of the time. The key difference here is that the bit error probability for N users talking all of the time is constant, whereas it is a *random* quantity (with the same mean) for $2N$ users talking half of the time.

In other words, asynchronous CDMA is ideally suited to a mobile network where large numbers of transmitters each generate a relatively small amount of traffic at irregular intervals. CDM (synchronous CDMA), TDMA, and FDMA systems cannot recover the underutilized resources inherent to bursty traffic due to the fixed number of orthogonal codes, time slots or frequency channels that can be assigned to individual transmitters. For instance, if there are N time slots in a TDMA system and $2N$ users that talk half of the time, then half of the time there will be more than N users needing to use more than N time slots. Furthermore, it would require significant overhead to continually allocate and deallocate the orthogonal-code, time-slot or frequency-channel resources. By comparison, asynchronous CDMA transmitters simply send when they have something to say and go off the air when they don't, keeping the same PN signature sequence as long as they are connected to the system.

Spread-spectrum Characteristics of CDMA

Most modulation schemes try to minimize the bandwidth of this signal since bandwidth is a limited resource. However, spread-spectrum techniques use a transmission bandwidth that is several orders of magnitude greater than the minimum required signal bandwidth. One of the initial reasons for doing this was military applications including guidance and communication systems. These systems were designed using spread spectrum because of its security and resistance to jamming. Asynchronous CDMA has some level of privacy built in because the signal is spread using a pseudo-random code; this code makes the spread-spectrum signals appear random or have noise-like properties. A receiver cannot demodulate this transmission without knowledge of the pseudo-random sequence used to encode the data. CDMA is also resistant to jamming. A jamming signal only has a finite amount of power available to jam the signal. The jammer can either spread its energy over the entire bandwidth of the signal or jam only part of the entire signal.

CDMA can also effectively reject narrow-band interference. Since narrow-band interference affects only a small portion of the spread-spectrum signal, it can easily be removed through notch filtering without much loss of information. Convolution encoding and interleaving can be used to assist in recovering this lost data. CDMA signals are also resistant to multipath fading. Since the spread-spectrum signal occupies a large bandwidth, only a small portion of this will undergo fading due to multipath at any given time. Like the narrow-band interference, this will result in only a small loss of data and can be overcome.

Another reason CDMA is resistant to multipath interference is because the delayed versions of the transmitted pseudo-random codes will have poor correlation with the original pseudo-random code, and will thus appear as another user, which is ignored at the receiver. In other words, as long as the multipath channel induces at least one chip of delay, the multipath signals will arrive at the receiver such that they are shifted in time by at least one chip from the intended signal. The correlation properties of the pseudo-random codes are such that this slight delay causes the multipath to appear uncorrelated with the intended signal, and it is thus ignored.

Some CDMA devices use a rake receiver, which exploits multipath delay components to improve the performance of the system. A rake receiver combines the information from several correlators, each one tuned to a different path delay, producing a stronger version of the signal than a simple receiver with a single correlation tuned to the path delay of the strongest signal.

Frequency reuse is the ability to reuse the same radio channel frequency at other cell sites within a cellular system. In the FDMA and TDMA systems, frequency planning is an important consideration. The frequencies used in different cells must be planned carefully to ensure signals from different cells do not interfere with each other. In a CDMA system, the same frequency can be used in every cell, because channelization is done using the pseudo-random codes. Reusing the same frequency in every cell eliminates the need for frequency planning in a CDMA system; however, planning of the different pseudo-random sequences must be done to ensure that the received signal from one cell does not correlate with the signal from a nearby cell.

Since adjacent cells use the same frequencies, CDMA systems have the ability to perform soft hand-offs. Soft hand-offs allow the mobile telephone to communicate simultaneously with two or more cells. The best signal quality is selected until the hand-off is complete. This is different from hard hand-offs utilized in other cellular systems. In a hard-hand-off situation, as the mobile telephone approaches a hand-off, signal strength may vary abruptly. In contrast, CDMA systems use the soft hand-off, which is undetectable and provides a more reliable and higher-quality signal.

Collaborative CDMA

In a recent study, a novel collaborative multi-user transmission and detection scheme called collaborative CDMA has been investigated for the uplink that exploits the differences between users' fading channel signatures to increase the user capacity well beyond the spreading length in the MAI-limited environment. The authors show that it is possible to achieve this increase at a low complexity and high bit error rate performance in flat fading channels, which is a major research challenge for overloaded CDMA systems. In this approach, instead of using one sequence per user as in conventional CDMA, the authors group a small number of users to share the same spreading sequence and enable group spreading and despreading operations. The new collaborative multi-user receiver consists of two stages: group multi-user detection (MUD) stage to suppress the MAI between the groups and a low-complexity maximum-likelihood detection stage to recover jointly the co-spread users' data using minimal Euclidean-distance measure and users' channel-gain coefficients. Further to note that research in the area is going on and in 2004, Prof. Li ping has introduced the new concept of enhanced CDMA version known as INTERLEAVE DIVISION MULTIPLE ACCESS (IDMA) scheme. It uses the orthogonal interleaved as the only means of user separation in place of signature sequence used in CDMA system.

GPRS

General Packet Radio Service (GPRS) is a packet oriented mobile data standard on the 2G and 3G cellular communication network's global system for mobile communications (GSM). GPRS was established by European Telecommunications Standards Institute (ETSI) in response to the earlier CDPD and i-mode packet-switched cellular technologies. It is now maintained by the 3rd Generation Partnership Project (3GPP).

GPRS is typically sold according to the total volume of data transferred during the billing cycle, in contrast with circuit switched data, which is usually billed per minute of connection time, or sometimes by one-third minute increments. Usage above the GPRS bundled data cap may be charged per MB of data, speed limited, or disallowed.

GPRS is a best-effort service, implying variable throughput and latency that depend on the number of other users sharing the service concurrently, as opposed to circuit switching, where a certain quality of service (QoS) is guaranteed during the connection. In 2G systems, GPRS provides data rates of 56–114 kbit/sec. 2G cellular technology combined with GPRS is sometimes described as *2.5G*, that is, a technology between the second (2G) and third (3G) generations of mobile telephony. It provides moderate-speed data transfer, by using unused time division multiple access (TDMA) channels in, for example, the GSM system. GPRS is integrated into GSM Release 97 and newer releases.

The GPRS core network allows 2G, 3G and WCDMA mobile networks to transmit IP packets to external networks such as the Internet. The GPRS system is an integrated part of the GSM network switching subsystem.

Services Offered

GPRS extends the GSM Packet circuit switched data capabilities and makes the following services possible:

- SMS messaging and broadcasting.

- "Always on" internet access.

- Multimedia messaging service (MMS).

- Push-to-talk over cellular (PoC).

- Instant messaging and presence—wireless village.

- Internet applications for smart devices through wireless application protocol (WAP).

- Point-to-point (P2P) service: inter-networking with the Internet (IP).

- Point-to-multipoint (P2M) service: point-to-multipoint multicast and point-to-multipoint group calls.

If SMS over GPRS is used, an SMS transmission speed of about 30 SMS messages per minute may be achieved. This is much faster than using the ordinary SMS over GSM, whose SMS transmission speed is about 6 to 10 SMS messages per minute.

Protocols Supported

GPRS supports the following protocols:

- Internet Protocol (IP). In practice, built-in mobile browsers use IPv4 before IPv6 is wide-spread.

- Point-to-Point Protocol (PPP) is typically not supported by mobile phone operators but if a cellular phone is used as a modem for a connected computer, PPP may be used to tunnel IP to the phone. This allows an IP address to be dynamically assigned (using IPCP rather than DHCP) to the mobile equipment.

- X.25 connections are typically used for applications like wireless payment terminals, although it has been removed from the standard. X.25 can still be supported over PPP, or even over IP, but this requires either a network-based router to perform encapsulation or software built into the end-device/terminal; e.g., user equipment (UE).

When TCP/IP is used, each phone can have one or more IP addresses allocated. GPRS will store and forward the IP packets to the phone even during handover. The TCP restores any packets lost (e.g. due to a radio noise induced pause).

Hardware

Devices supporting GPRS are grouped into three classes:

- Class A: Can be connected to GPRS service and GSM service (voice, SMS) simultaneously. Such devices are now available.

- Class B: Can be connected to GPRS service and GSM service (voice, SMS), but using only one at a time. During GSM service (voice call or SMS), GPRS service is suspended and resumed automatically after the GSM service (voice call or SMS) has concluded. Most GPRS mobile devices are Class B.

- Class C: Are connected to either GPRS service or GSM service (voice, SMS) and must be switched manually between one service and the other.

Because a Class A device must service GPRS and GSM networks together, it effectively needs two radios. To avoid this hardware requirement, a GPRS mobile device may implement the dual transfer mode (DTM) feature. A DTM-capable mobile can handle both GSM packets and GPRS packets with network coordination to ensure both types are not transmitted at the same time. Such devices are considered pseudo-Class A, sometimes referred to as "simple class A". Some networks have supported DTM since 2007.

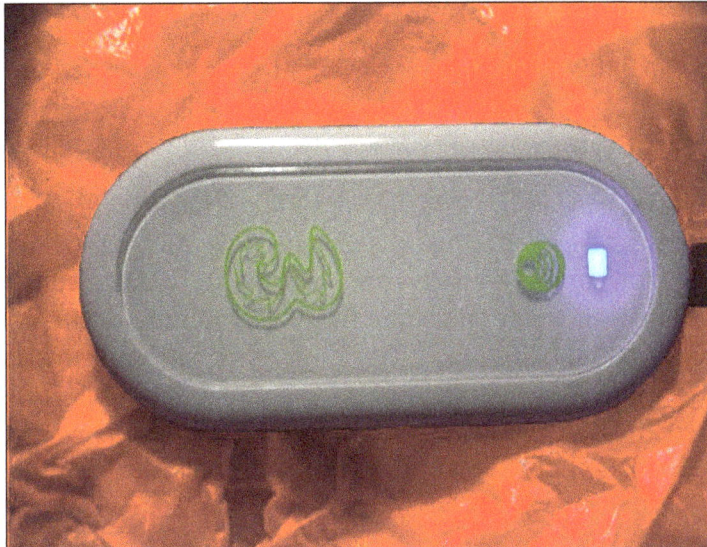

Huawei E220 3G/GPRS Modem.

USB 3G/GPRS modems have a terminal-like interface over USB with V.42bis, and RFC 1144 data formats. Some models include an external antenna connector. Modem cards for laptop PCs, or external USB modems are available, similar in shape and size to a computer mouse, or a pendrive.

Addressing

A GPRS connection is established by reference to its access point name (APN). The APN defines the services such as wireless application protocol (WAP) access, short message service (SMS), multimedia messaging service (MMS), and for Internet communication services such as email and World Wide Web access.

In order to set up a GPRS connection for a wireless modem, a user must specify an APN, optionally a user name and password, and very rarely an IP address, provided by the network operator.

GPRs Modems and Modules

GSM module or GPRS modules are similar to modems, but there's one difference: the modem is an external piece of equipment, whereas the GSM module or GPRS module can be integrated within electrical or electronic equipment. It is an embedded piece of hardware. A GSM mobile, on the other hand, is a complete embedded system in itself. It comes with embedded processors dedicated to provide a functional interface between the user and the mobile network.

Coding Schemes and Speeds

The upload and download speeds that can be achieved in GPRS depend on a number of factors such as:

- The number of BTS TDMA time slots assigned by the operator.

- The channel encoding used.

- The maximum capability of the mobile device expressed as a GPRS multislot class.

Multiple Access Schemes

The multiple access methods used in GSM with GPRS are based on frequency division duplex (FDD) and TDMA. During a session, a user is assigned to one pair of up-link and down-link frequency channels. This is combined with time domain statistical multiplexing which makes it possible for several users to share the same frequency channel. The packets have constant length, corresponding to a GSM time slot. The down-link uses first-come first-served packet scheduling, while the up-link uses a scheme very similar to reservation ALOHA (R-ALOHA). This means that slotted ALOHA (S-ALOHA) is used for reservation inquiries during a contention phase, and then the actual data is transferred using dynamic TDMA with first-come first-served.

Channel Encoding

The channel encoding process in GPRS consists of two steps: first, a cyclic code is used to add parity bits, which are also referred to as the Block Check Sequence, followed by coding with a possibly punctured convolutional code. The Coding Schemes CS-1 to CS-4 specifies the number of parity bits generated by the cyclic code and the puncturing rate of the convolutional code. In Coding Schemes CS-1 through CS-3, the convolutional code is of rate 1/2, i.e. each input bit is converted into two coded bits. In Coding Schemes CS-2 and CS-3, the output of the convolutional code is punctured to achieve the desired code rate. In Coding Scheme CS-4, no convolutional coding is applied. The following table summarises the options.

GPRS Coding scheme	Bitrate including RLC/MAC overhead (kbit/s/slot)	Bitrate excluding RLC/MAC overhead (kbit/s/slot)	Modulation	Code rate
CS-1	9.20	8.00	GMSK	1/2
CS-2	13.55	12.00	GMSK	≈2/3
CS-3	15.75	14.40	GMSK	≈3/4
CS-4	21.55	20.00	GMSK	1

The least robust, but fastest, coding scheme (CS-4) is available near a base transceiver station (BTS), while the most robust coding scheme (CS-1) is used when the mobile station (MS) is further away from a BTS.

Using the CS-4 it is possible to achieve a user speed of 20.0 kbit/s per time slot. However, using this scheme the cell coverage is 25% of normal. CS-1 can achieve a user speed of only 8.0 kbit/s per time slot, but has 98% of normal coverage. Newer network equipment can adapt the transfer speed automatically depending on the mobile location.

In addition to GPRS, there are two other GSM technologies which deliver data services: circuit-switched data (CSD) and high-speed circuit-switched data (HSCSD). In contrast to the shared nature of GPRS, these instead establish a dedicated circuit (usually billed per minute). Some applications such as video calling may prefer HSCSD, especially when there is a continuous flow of data between the endpoints.

The following table summarises some possible configurations of GPRS and circuit switched data services.

Technology	Download (kbit/s)	Upload (kbit/s)	TDMA timeslots allocated (DL+UL)
CSD	9.6	9.6	1+1
HSCSD	28.8	14.4	2+1
HSCSD	43.2	14.4	3+1
GPRS	85.6	21.4 (Class 8 & 10 and CS-4)	4+1
GPRS	64.2	42.8 (Class 10 and CS-4)	3+2
EGPRS (EDGE)	236.8	59.2 (Class 8, 10 and MCS-9)	4+1
EGPRS (EDGE)	177.6	118.4 (Class 10 and MCS-9)	3+2

Multislot Class

The multislot class determines the speed of data transfer available in the Uplink and Downlink directions. It is a value between 1 and 45 which the network uses to allocate radio channels in the uplink and downlink direction. Multislot class with values greater than 31 are referred to as high multislot classes.

A multislot allocation is represented as, for example, 5+2. The first number is the number of downlink timeslots and the second is the number of uplink timeslots allocated for use by the mobile station. A commonly used value is class 10 for many GPRS/EGPRS mobiles which uses a maximum of 4 timeslots in downlink direction and 2 timeslots in uplink direction. However simultaneously a maximum number of 5 simultaneous timeslots can be used in both uplink and downlink. The network will automatically configure for either 3+2 or 4+1 operation depending on the nature of data transfer.

Some high end mobiles, usually also supporting UMTS, also support GPRS/EDGE multislot class 32. According to 3GPP TS 45.002 (Release 12), Table B.1, mobile stations of this class support 5 timeslots in downlink and 3 timeslots in uplink with a maximum number of 6 simultaneously used timeslots. If data traffic is concentrated in downlink direction the network will configure the connection for 5+1 operation. When more data is transferred in the uplink the network can at any time change the constellation to 4+2 or 3+3. Under the best reception conditions, i.e. when the best EDGE modulation and coding scheme can be used, 5 timeslots can carry a bandwidth of 5*59.2 kbit/s = 296 kbit/s. In uplink direction, 3 timeslots can carry a bandwidth of 3*59.2 kbit/s = 177.6 kbit/s.

Multislot Classes for GPRS/EGPRS

Multislot Class	Downlink TS	Uplink TS	Active TS
1	1	1	2
2	2	1	3
3	2	2	3
4	3	1	4
5	2	2	4
6	3	2	4
7	3	3	4
8	4	1	5
9	3	2	5
10	4	2	5
11	4	3	5
12	4	4	5
30	5	1	6
31	5	2	6
32	5	3	6
33	5	4	6
34	5	5	6

Attributes of a Multislot Class

Each multislot class identifies the following:

- The maximum number of Timeslots that can be allocated on uplink.

- The maximum number of Timeslots that can be allocated on downlink.

- The total number of timeslots which can be allocated by the network to the mobile.

- The time needed for the MS to perform adjacent cell signal level measurement and get ready to transmit.

- The time needed for the MS to get ready to transmit.

- The time needed for the MS to perform adjacent cell signal level measurement and get ready to receive.

- The time needed for the MS to get ready to receive.

The different multislot class specification is detailed in the Annex B of the 3GPP Technical Specification 45.002 (Multiplexing and multiple access on the radio path).

Usability

The maximum speed of a GPRS connection offered in 2003 was similar to a modem connection in an analog wire telephone network, about 32–40 kbit/s, depending on the phone used. Latency is very high; round-trip time (RTT) is typically about 600–700 ms and often reaches 1s. GPRS is typically prioritized lower than speech, and thus the quality of connection varies greatly.

Devices with latency/RTT improvements (via, for example, the extended UL TBF mode feature) are generally available. Also, network upgrades of features are available with certain operators. With these enhancements the active round-trip time can be reduced, resulting in significant increase in application-level throughput speeds.

FDMA

FDMA (stands for Frequency Division Multiple Access) is the signal multiplexing technology used in the Advanced Mobile Phone Service (AMPS) analog version of cellular phone technology.

The signal multiplexing technology used in the Advanced Mobile Phone Service (AMPS) analog version of cellular phone technology. Frequency Division Multiple Access (FDMA) is one of three methods used for allocating channels to users over the shared wireless communications medium in cellular phone communication; the others are Time Division Multiple Access (TDMA) and Code Division Multiple Access (CDMA).

How FDMA Works?

FDMA is implemented at the media access control (MAC) layer of the data-link layer in the Open Systems Interconnection (OSI) reference model for networking protocol stacks. FDMA is based on the frequency-division multiplexing (FDM) technique used in wireless networking. In FDMA, the user is assigned a specific frequency band in the electromagnetic spectrum, and during a call that user is the only one who has the right to access the specific band. In the AMPS cellular phone system, these frequency bands are allocated from the electromagnetic spectrum as follows:

- Transmission by mobile station: 824 MHz to 849 MHz.

- Transmission by base station: 869 MHz to 894 MHz.

Two different frequency bands are used to allow full-duplex communication between base and mobile stations. Both of these bands are then divided into discrete channels that are 30 kHz wide in bandwidth.

OFDMA

Orthogonal frequency-division multiple access (OFDMA) is a multi-user version of the popular orthogonal frequency-division multiplexing (OFDM) digital modulation scheme. Multiple access is achieved in OFDMA by assigning subsets of subcarriers to individual users. This allows simultaneous low-data-rate transmission from several users.

Claimed Advantages Over OFDM with Time-domain Statistical Multiplexing

- Allows simultaneous low-data-rate transmission from several users.
- Pulsed carrier can be avoided.
- Lower maximal transmission power for low-data-rate users.
- Shorter delay and constant delay.
- Contention-based multiple access (collision avoidance) is simplified.
- Further improves OFDM robustness to fading and interference.
- Combat narrow-band interference.

Claimed OFDMA Advantages

- Flexibility of deployment across various frequency bands with little needed modification to the air interface.
- Averaging interferences from neighbouring cells, by using different basic carrier permutations between users in different cells.
- Interferences within the cell are averaged by using allocation with cyclic permutations.
- Enables single-frequency network coverage, where coverage problem exists and gives excellent coverage.
- Offers frequency diversity by spreading the carriers all over the used spectrum.
- Allows per-channel or per-subchannel power.

Recognised Disadvantages of OFDMA

- Higher sensitivity to frequency offsets and phase noise.
- Asynchronous data communication services such as web access are characterised by short communication bursts at high data rate. Few users in a base station cell are transferring data simultaneously at low constant data rate.
- The complex OFDM electronics, including the FFT algorithm and forward error correction, are constantly active independent of the data rate, which is inefficient from power-consumption point of view, while OFDM combined with data packet scheduling may allow FFT algorithm to hibernate during certain time intervals.
- The OFDM diversity gain and resistance to frequency-selective fading may partly be lost if very few sub-carriers are assigned to each user, and if the same carrier is used in every OFDM symbol. Adaptive sub-carrier assignment based on fast feedback information about the channel, or sub-carrier frequency hopping, is therefore desirable.
- Dealing with co-channel interference from nearby cells is more complex in OFDM than in

CDMA. It would require dynamic channel allocation with advanced coordination among adjacent base stations.

- The fast channel feedback information and adaptive sub-carrier assignment is more complex than CDMA fast power control.

Characteristics and Principles of Operation

Based on feedback information about the channel conditions, adaptive user-to-subcarrier assignment can be achieved. If the assignment is done sufficiently fast, this further improves the OFDM robustness to fast fading and narrow-band cochannel interference, and makes it possible to achieve even better system spectral efficiency.

Different numbers of sub-carriers can be assigned to different users, in view to support differentiated Quality of Service (QoS), i.e. to control the data rate and error probability individually for each user.

OFDMA can be seen as an alternative to combining OFDM with time-division multiple access (TDMA) or time-domain statistical multiplexing communication. Low-data-rate users can send continuously with low transmission power instead of using a "pulsed" high-power carrier. Constant delay, and shorter delay, can be achieved.

OFDMA can also be described as a combination of frequency-domain and time-domain multiple access, where the resources are partitioned in the time–frequency space, and slots are assigned along the OFDM symbol index, as well as OFDM sub-carrier index.

OFDMA is considered as highly suitable for broadband wireless networks, due to advantages including scalability and use of multiple antennas (MIMO)-friendliness, and ability to take advantage of channel frequency selectivity.

In spectrum sensing cognitive radio, OFDMA is a possible approach to filling free radio frequency bands adaptively. Timo A. Weiss and Friedrich K. Jondral of the University of Karlsruhe proposed a spectrum pooling system in which free bands sensed by nodes were immediately filled by OFDMA subbands.

Usage

OFDMA is used in:

- The mobility mode of the IEEE 802.16 Wireless MAN standard, commonly referred to as wimax.

- The wireless LAN (WLAN) standard IEEE 802.11ax.

- The IEEE 802.20 mobile Wireless MAN standard, commonly referred to as MBWA.

- Moca 2.0.

- The downlink of the 3GPP Long-Term Evolution (LTE) fourth-generation mobile broadband standard. The radio interface was formerly named *High Speed OFDM Packet Access* (HSOPA), now named Evolved UMTS Terrestrial Radio Access (E-UTRA).

- The Qualcomm Flarion Technologies Mobile Flash-OFDM.

- The now defunct Qualcomm/3GPP2 Ultra Mobile Broadband (UMB) project, intended as a successor of CDMA2000, but replaced by LTE.

OFDMA is also a candidate access method for the IEEE 802.22 *Wireless Regional Area Networks* (WRAN), a cognitive radio technology which uses white spaces in the television (TV) frequency spectrum, and the proposed access method for DECT-5G specification which aims to fulfill IMT-2020 requirements for high-throughput mobile broadband (eMMB) and ultra-reliable low latency(URLLC) applications.

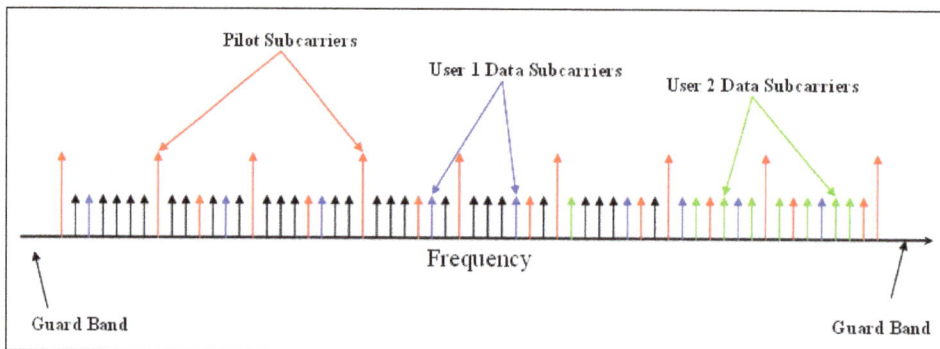

OFDMA subcarriers.

SDMA

Spatial division multiple access (SDMA) is a channel access method used in mobile communication systems which reuses the same set of cell phone frequencies in a given service area. Two cells or two small regions can make use of the same set of frequencies if they are separated by an allowable distance (called the reuse distance).

SDMA increases the capacity of the system and transmission quality by focusing the signal into narrow transmission beams. Through the use of smart antennas with beams pointed at the direction of the mobile station, SDMA serves different users within the same region.

Mobile stations operating outside the bounds of these directed beams experience a near zero interference from other mobile stations operating under the same base station with the same radio frequency.

Since the beams are focused, the radio energy frequency can have increased base station range. This attribute of SDMA allows base stations to have larger radio coverage with less radiated energy. This narrow beam width also allows greater gain and clarity.

Under traditional mobile phone network systems, the base station radiates radio signals in all directions within the cell without knowledge of the location of the mobile station. SDMA technology channels radio signals based on the location of the mobile station. Through this method, the SDMA architecture saves on valuable network resources and prevents redundant signal transmission in areas where mobile devices are currently inactive.

The main advantage of SDMA is frequency reuse. Provided the reuse distance is preserved in the network architecture, interference can be near zero, even if mobile stations use the same allocated frequencies.

LTE

In telecommunication, Long-Term Evolution (LTE) is a standard for wireless broadband communication for mobile devices and data terminals, based on the GSM/EDGE and UMTS/HSPA technologies. It increases the capacity and speed using a different radio interface together with core network improvements. The standard is developed by the 3GPP (3rd Generation Partnership Project) and is specified in its Release 8 document series, with minor enhancements described in Release 9. LTE is the upgrade path for carriers with both GSM/UMTS networks and CDMA2000 networks. The different LTE frequencies and bands used in different countries mean that only multi-band phones are able to use LTE in all countries where it is supported.

LTE is commonly marketed as "4G LTE and Advance 4G", but it does not meet the technical criteria of a 4G wireless service, as specified in the 3GPP Release 8 and 9 document series for LTE Advanced. LTE is also commonly known as 3.95G. The requirements were originally set forth by the ITU-R organization in the IMT Advanced specification. However, due to marketing pressures and the significant advancements that WiMAX, Evolved High Speed Packet Access and LTE bring to the original 3G technologies, ITU later decided that LTE together with the aforementioned technologies can be called 4G technologies. The LTE Advanced standard formally satisfies the ITU-R requirements to be considered IMT-Advanced. To differentiate LTE Advanced and WiMAX-Advanced from current 4G technologies, ITU has defined them as "True 4G".

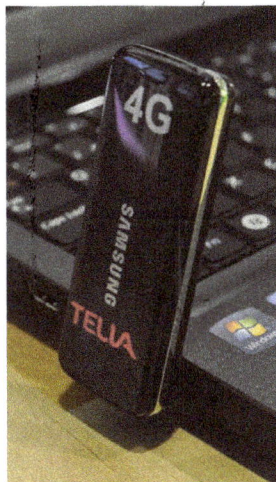

Telia-branded Samsung LTE modem.

LTE stands for Long Term Evolution and is a registered trademark owned by ETSI (European Telecommunications Standards Institute) for the wireless data communications technology and a development of the GSM/UMTS standards. However, other nations and companies do play an active role in the LTE project. The goal of LTE was to increase the capacity and speed of wireless data networks using new DSP (digital signal processing) techniques and modulations that were developed around the turn of the millennium. A further goal was the redesign and simplification of the network architecture to an IP-based system with significantly reduced transfer latency compared to the 3G architecture. The LTE wireless interface is incompatible with 2G and 3G networks, so that it must be operated on a separate radio spectrum.

LTE was first proposed in 2004 by Japan's NTT Docomo, with studies on the standard officially commenced in 2005. In May 2007, the LTE/SAE Trial Initiative (LSTI) alliance was founded as a global collaboration between vendors and operators with the goal of verifying and promoting the new standard in order to ensure the global introduction of the technology as quickly as possible. The LTE standard was finalized in December 2008, and the first publicly available LTE service was launched by TeliaSonera in Oslo and Stockholm on December 14, 2009, as a data connection with a USB modem. The LTE services were launched by major North American carriers as well, with the Samsung SCH-r900 being the world's first LTE Mobile phone starting on September 21, 2010, and Samsung Galaxy Indulge being the world's first LTE smartphone starting on February 10, 2011, both offered by MetroPCS, and the HTC ThunderBolt offered by Verizon starting on March 17 being the second LTE smartphone to be sold commercially. In Canada, Rogers Wireless was the first to launch LTE network on July 7, 2011, offering the Sierra Wireless AirCard 313U USB mobile broadband modem, known as the "LTE Rocket stick" then followed closely by mobile devices from both HTC and Samsung. Initially, CDMA operators planned to upgrade to rival standards called UMB and WiMAX, but major CDMA operators (such as Verizon, Sprint and MetroPCS in the United States, Bell and Telus in Canada, au by KDDI in Japan, SK Telecom in South Korea and China Telecom/China Unicom in China) have announced instead they intend to migrate to LTE. The next version of LTE is LTE Advanced, which was standardized in March 2011. Services are expected to commence in 2013. Additional evolution known as LTE Advanced Pro have been approved in year 2015.

HTC ThunderBolt, the second commercially available LTE smartphone.

The LTE specification provides downlink peak rates of 300 Mbit/s, uplink peak rates of 75 Mbit/s and QoS provisions permitting a transfer latency of less than 5 ms in the radio access network. LTE has the ability to manage fast-moving mobiles and supports multi-cast and broadcast streams. LTE supports scalable carrier bandwidths, from 1.4 MHz to 20 MHz and supports both frequency division duplexing (FDD) and time-division duplexing (TDD). The IP-based network architecture, called the Evolved Packet Core (EPC) designed to replace the GPRS Core Network, supports seamless handovers for both voice and data to cell towers with older network technology such as GSM, UMTS and CDMA2000. The simpler architecture results in lower operating costs (for example, each E-UTRA cell will support up to four times the data and voice capacity supported by HSPA).

LTE-TDD and LTE-FDD

Long-Term Evolution Time-Division Duplex (LTE-TDD), also referred to as TDD LTE, is a 4G telecommunications technology and standard co-developed by an international coalition of companies, including China Mobile, Datang Telecom, Huawei, ZTE, Nokia Solutions and Networks, Qualcomm, Samsung, and ST-Ericsson. It is one of the two mobile data transmission technologies of the Long-Term Evolution (LTE) technology standard, the other being Long-Term Evolution Frequency-Division Duplex (LTE-FDD). While some companies refer to LTE-TDD as "TD-LTE" for familiarity with TD-SCDMA, there is no reference to that acronym anywhere in the 3GPP specifications.

There are two major differences between LTE-TDD and LTE-FDD: how data is uploaded and downloaded, and what frequency spectra the networks are deployed in. While LTE-FDD uses paired frequencies to upload and download data, LTE-TDD uses a single frequency, alternating between uploading and downloading data through time. The ratio between uploads and downloads on a LTE-TDD network can be changed dynamically, depending on whether more data needs to be sent or received. LTE-TDD and LTE-FDD also operate on different frequency bands, with LTE-TDD working better at higher frequencies, and LTE-FDD working better at lower frequencies. Frequencies used for LTE-TDD range from 1850 MHz to 3800 MHz, with several different bands being used. The LTE-TDD spectrum is generally cheaper to access, and has less traffic. Further, the bands for LTE-TDD overlap with those used for WiMAX, which can easily be upgraded to support LTE-TDD.

Despite the differences in how the two types of LTE handle data transmission, LTE-TDD and LTE-FDD share 90 percent of their core technology, making it possible for the same chipsets and networks to use both versions of LTE. A number of companies produce dual-mode chips or mobile devices, including Samsung and Qualcomm, while operators CMHK and Hi3G Access have developed dual-mode networks in Hong Kong and Sweden, respectively.

The creation of LTE-TDD involved a coalition of international companies that worked to develop and test the technology. China Mobile was an early proponent of LTE-TDD, along with other companies like Datang Telecom and Huawei, which worked to deploy LTE-TDD networks, and later developed technology allowing LTE-TDD equipment to operate in white spaces—frequency spectra between broadcast TV stations. Intel also participated in the development, setting up a LTE-TDD interoperability lab with Huawei in China, as well as ST-Ericsson, Nokia, and Nokia Siemens, which developed LTE-TDD base stations that increased capacity by 80 percent and coverage by 40 percent. Qualcomm also participated, developing the world's first multi-mode chip, combining both LTE-TDD and LTE-FDD, along with HSPA and EV-DO. Accelleran, a Belgian company, has also worked to build small cells for LTE-TDD networks.

Trials of LTE-TDD technology began as early as 2010, with Reliance Industries and Ericsson India conducting field tests of LTE-TDD in India, achieving 80 megabit-per second download speeds and 20 megabit-per-second upload speeds. By 2011, China Mobile began trials of the technology in six cities.

Although initially seen as a technology utilized by only a few countries, including China and India, by 2011 international interest in LTE-TDD had expanded, especially in Asia, in part due to LTE-TDD 's lower cost of deployment compared to LTE-FDD. By the middle of that year, 26 networks around the world were conducting trials of the technology. The Global LTE-TDD Initiative (GTI) was also started in 2011, with founding partners China Mobile, Bharti Airtel, SoftBank Mobile, Vodafone, Clearwire, Aero2 and E-Plus. In September 2011, Huawei announced it would partner with Polish mobile provider Aero2 to develop a combined LTE-TDD and LTE-FDD network in Poland, and by April 2012, ZTE Corporation had worked to deploy trial or commercial LTE-TDD networks for 33 operators in 19 countries. In late 2012, Qualcomm worked extensively to deploy a commercial LTE-TDD network in India, and partnered with Bharti Airtel and Huawei to develop the first multi-mode LTE-TDD smartphone for India.

In Japan, SoftBank Mobile launched LTE-TDD services in February 2012 under the name Advanced eXtended Global Platform (AXGP), and marketed as SoftBank 4G (ja). The AXGP band was previously used for Willcom's PHS service, and after PHS was discontinued in 2010 the PHS band was re-purposed for AXGP service.

In the U.S., Clearwire planned to implement LTE-TDD, with chip-maker Qualcomm agreeing to support Clearwire's frequencies on its multi-mode LTE chipsets. With Sprint's acquisition of Clearwire in 2013, the carrier began using these frequencies for LTE service on networks built by Samsung, Alcatel-Lucent, and Nokia.

As of March 2013, 156 commercial 4G LTE networks existed, including 142 LTE-FDD networks and 14 LTE-TDD networks. As of November 2013, the South Korean government planned to allow a fourth wireless carrier in 2014, which would provide LTE-TDD services, and in December 2013, LTE-TDD licenses were granted to China's three mobile operators, allowing commercial deployment of 4G LTE services.

In January 2014, Nokia Solutions and Networks indicated that it had completed a series of tests of voice over LTE (VoLTE) calls on China Mobile's TD-LTE network. The next month, Nokia Solutions and Networks and Sprint announced that they had demonstrated throughput speeds of 2.6 gigabits per second using a LTE-TDD network, surpassing the previous record of 1.6 gigabits per second.

Features

Much of the LTE standard addresses the upgrading of 3G UMTS to what will eventually be 4G mobile communications technology. A large amount of the work is aimed at simplifying the architecture of the system, as it transitions from the existing UMTS circuit + packet switching combined network, to an all-IP flat architecture system. E-UTRA is the air interface of LTE. Its main features are:

- Peak download rates up to 299.6 Mbit/s and upload rates up to 75.4 Mbit/s depending on the user equipment category (with 4×4 antennas using 20 MHz of spectrum). Five different terminal classes have been defined from a voice-centric class up to a high-end terminal that supports the peak data rates. All terminals will be able to process 20 MHz bandwidth.

- Low data transfer latencies (sub-5 ms latency for small IP packets in optimal conditions), lower latencies for handover and connection setup time than with previous radio access technologies.

- Improved support for mobility, exemplified by support for terminals moving at up to 350 km/h (220 mph) or 500 km/h (310 mph) depending on the frequency band.

- Orthogonal frequency-division multiple access for the downlink, Single-carrier FDMA for the uplink to conserve power.

- Support for both FDD and TDD communication systems as well as half-duplex FDD with the same radio access technology.

- Support for all frequency bands currently used by IMT systems by ITU-R.

- Increased spectrum flexibility: 1.4 MHz, 3 MHz, 5 MHz, 10 MHz, 15 MHz and 20 MHz wide cells are standardized. (W-CDMA has no option for other than 5 MHz slices, leading to some problems rolling-out in countries where 5 MHz is a commonly allocated width of spectrum so would frequently already be in use with legacy standards such as 2G GSM and cdmaOne).

- Support for cell sizes from tens of metres radius (femto and picocells) up to 100 km (62 miles) radius macrocells. In the lower frequency bands to be used in rural areas, 5 km (3.1 miles) is the optimal cell size, 30 km (19 miles) having reasonable performance, and up to 100 km cell sizes supported with acceptable performance. In the city and urban areas, higher frequency bands (such as 2.6 GHz in EU) are used to support high-speed mobile broadband. In this case, cell sizes may be 1 km (0.62 miles) or even less.

- Support of at least 200 active data clients in every 5 MHz cell.

- Simplified architecture: The network side of E-UTRAN is composed only of eNode Bs.

- Support for inter-operation and co-existence with legacy standards (e.g., GSM/EDGE, UMTS and CDMA2000). Users can start a call or transfer of data in an area using an LTE standard and should coverage be unavailable, continue the operation without any action on their part using GSM/GPRS or W-CDMA-based UMTS or even 3GPP2 networks such as cdmaOne or CDMA2000.

- Uplink and downlink Carrier aggregation.

- Packet-switched radio interface.

- Support for MBSFN (multicast-broadcast single-frequency network). This feature can deliver services such as Mobile TV using the LTE infrastructure, and is a competitor for DVB-H-based TV broadcast only LTE compatible devices receives LTE signal.

Voice Calls

The LTE standard supports only packet switching with its all-IP network. Voice calls in GSM, UMTS and CDMA2000 are circuit switched, so with the adoption of LTE, carriers will have to re-engineer their voice call network. Three different approaches sprang up.

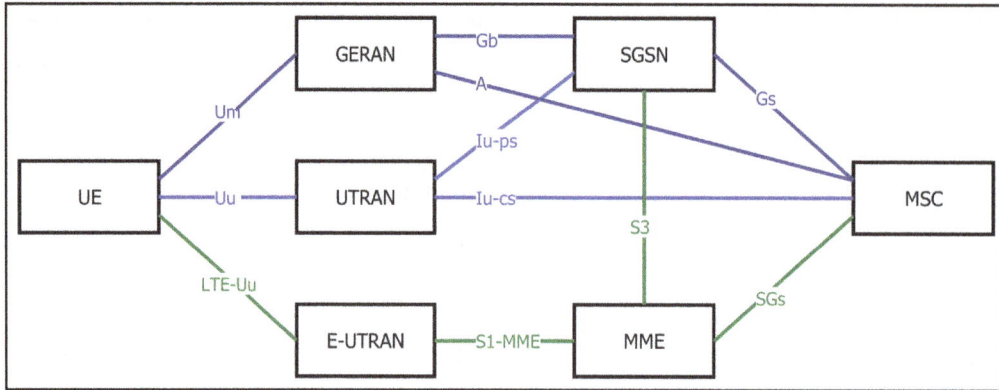

CS domLTE CSFB to GSM/UMTS network interconnects.

Voice Over LTE (VoLTE)

Circuit-switched Fallback (CSFB)

In this approach, LTE just provides data services, and when a voice call is to be initiated or received, it will fall back to the circuit-switched domain. When using this solution, operators just need to upgrade the MSC instead of deploying the IMS, and therefore, can provide services quickly. However, the disadvantage is longer call setup delay.

Simultaneous Voice and LTE (SVLTE)

In this approach, the handset works simultaneously in the LTE and circuit switched modes, with the LTE mode providing data services and the circuit switched mode providing the voice service. This is a solution solely based on the handset, which does not have special requirements on the network and does not require the deployment of IMS either. The disadvantage of this solution is that the phone can become expensive with high power consumption.

Single Radio Voice Call Continuity (SRVCC)

One additional approach which is not initiated by operators is the usage of over-the-top content (OTT) services, using applications like Skype and Google Talk to provide LTE voice service.

Most major backers of LTE preferred and promoted VoLTE from the beginning. The lack of software support in initial LTE devices, as well as core network devices, however led to a number of carriers promoting VoLGA (Voice over LTE Generic Access) as an interim solution. The idea was to use the same principles as GAN (Generic Access Network, also known as UMA or Unlicensed Mobile Access), which defines the protocols through which a mobile handset can perform voice calls over a customer's private Internet connection, usually over wireless LAN. VoLGA however never gained much support, because VoLTE (IMS) promises much more flexible services, albeit at the cost of having to upgrade the entire voice call infrastructure. VoLTE will also require Single Radio Voice Call Continuity (SRVCC) in order to be able to smoothly perform a handover to a 3G network in case of poor LTE signal quality.

While the industry has seemingly standardized on VoLTE for the future, the demand for voice calls today has led LTE carriers to introduce circuit-switched fallback as a stopgap measure. When placing or receiving a voice call, LTE handsets will fall back to old 2G or 3G networks for the duration of the call.

Enhanced Voice Quality

To ensure compatibility, 3GPP demands at least AMR-NB codec (narrow band), but the recommended speech codec for VoLTE is Adaptive Multi-Rate Wideband, also known as HD Voice. This codec is mandated in 3GPP networks that support 16 kHz sampling.

Fraunhofer IIS has proposed and demonstrated "Full-HD Voice", an implementation of the AAC-ELD (Advanced Audio Coding – Enhanced Low Delay) codec for LTE handsets. Where previous cell phone voice codecs only supported frequencies up to 3.5 kHz and upcoming wideband audio services branded as *HD Voice* up to 7 kHz, Full-HD Voice supports the entire bandwidth range from 20 Hz to 20 kHz. For end-to-end Full-HD Voice calls to succeed, however, both the caller and recipient's handsets, as well as networks, have to support the feature.

Patents

According to the European Telecommunications Standards Institute's (ETSI) intellectual property rights (IPR) database, about 50 companies have declared, as of March 2012, holding essential patents covering the LTE standard. The ETSI has made no investigation on the correctness of the declarations however, so that "any analysis of essential LTE patents should take into account more than ETSI declarations." Independent studies have found that about 3.3 to 5 percent of all revenues from handset manufacturers are spent on standard-essential patents. This is less than the combined published rates, due to reduced-rate licensing agreements, such as cross-licensing.

LTE Advanced

LTE Advanced logo.

LTE Advanced is a mobile communication standard and a major enhancement of the Long Term Evolution (LTE) standard. It was formally submitted as a candidate 4G to ITU-T in late 2009 as meeting the requirements of the IMT-Advanced standard, and was standardized by the 3rd Generation Partnership Project (3GPP) in March 2011 as 3GPP Release 10.

The LTE format was first proposed by NTT DoCoMo of Japan and has been adopted as the international standard. LTE standardization has matured to a state where changes in the specification are limited to corrections and bug fixes. The first commercial services were launched in Sweden and Norway in December 2009 followed by the United States and Japan in 2010. More LTE networks were deployed globally during 2010 as a natural evolution of several 2G and 3G systems, including Global system for mobile communications (GSM) and Universal Mobile Telecommunications System (UMTS) in the 3GPP family as well as CDMA2000 in the 3GPP2 family.

LTE Advanced (with carrier aggregation) signal indicator in Android.

The work by 3GPP to define a 4G candidate radio interface technology started in Release 9 with the study phase for LTE-Advanced. Being described as a 3.9G (beyond 3G but pre-4G), the first release of LTE did not meet the requirements for 4G (also called IMT Advanced as defined by the International Telecommunication Union) such as peak data rates up to 1 Gb/s. The ITU has invited the submission of candidate Radio Interface Technologies (RITs) following their requirements in a circular letter, 3GPP Technical Report (TR) 36.913, "Requirements for Further Advancements for E-UTRA (LTE-Advanced)." These are based on ITU's requirements for 4G and on operators' own requirements for advanced LTE. Major technical considerations include the following:

- Continual improvement to the LTE radio technology and architecture.

- Scenarios and performance requirements for working with legacy radio technologies.

- Backward compatibility of LTE-Advanced with LTE. An LTE terminal should be able to work in an LTE-Advanced network and vice versa. Any exceptions will be considered by 3GPP.

- Consideration of recent World Radiocommunication Conference (WRC-07) decisions regarding frequency bands to ensure that LTE-Advanced accommodates the geographically available spectrum for channels above 20 MHz. Also, specifications must recognize those parts of the world in which wideband channels are not available.

Likewise, 'WiMAX 2', 802.16m, has been approved by ITU as the IMT Advanced family. WiMAX 2 is designed to be backward compatible with WiMAX 1 devices. Most vendors now support conversion of 'pre-4G', pre-advanced versions and some support software upgrades of base station equipment from 3G.

The mobile communication industry and standards organizations have therefore started work on 4G access technologies, such as LTE Advanced. At a workshop in April 2008 in China, 3GPP agreed the plans for work on Long Term Evolution (LTE). A first set of specifications were approved in June 2008. Besides the peak data rate 1 Gb/s as defined by the ITU-R, it also targets faster switching between power states and improved performance at the cell edge. Detailed proposals are being studied within the working groups.

Three technologies from the LTE-Advanced tool-kit – carrier aggregation, 4x4 MIMO and 256QAM modulation in the downlink – if used together and with sufficient aggregated bandwidth, can deliver maximum peak downlink speeds approaching, or even exceeding, 1 Gbit/s. Such networks are often described as 'Gigabit LTE networks' mirroring a term that is also used in the fixed broadband industry.

Proposals

The target of 3GPP LTE Advanced is to reach and surpass the ITU requirements. LTE Advanced should be compatible with first release LTE equipment, and should share frequency bands with first release LTE. In the feasibility study for LTE Advanced, 3GPP determined that LTE Advanced would meet the ITU-R requirements for 4G. The results of the study are published in 3GPP Technical Report (TR) 36.912.

One of the important LTE Advanced benefits is the ability to take advantage of advanced topology networks; optimized heterogeneous networks with a mix of macrocells with low power nodes such as picocells, femtocells and new relay nodes. The next significant performance leap in wireless networks will come from making the most of topology, and brings the network closer to the user by adding many of these low power nodes — LTE Advanced further improves the capacity and coverage, and ensures user fairness. LTE Advanced also introduces multicarrier to be able to use ultra wide bandwidth, up to 100 MHz of spectrum supporting very high data rates.

In the research phase many proposals have been studied as candidates for LTE Advanced (LTE-A) technologies. The proposals could roughly be categorized into:

- Support for relay node base stations.

- Coordinated multipoint (CoMP) transmission and reception.

- UE Dual TX antenna solutions for SU-MIMO and diversity MIMO, commonly referred to as 2x2 MIMO.

- Scalable system bandwidth exceeding 20 MHz, up to 100 MHz.

- Carrier aggregation of contiguous and non-contiguous spectrum allocations.

- Local area optimization of air interface.

- Nomadic/Local Area network and mobility solutions.

- Flexible spectrum usage.

- Cognitive radio.

- Automatic and autonomous network configuration and operation.

- Support of autonomous network and device test, measurement tied to network management and optimization.

- Enhanced precoding and forward error correction.

- Interference management and suppression.

- Asymmetric bandwidth assignment for FDD.

- Hybrid OFDMA and SC-FDMA in uplink.

- UL/DL inter eNB coordinated MIMO.

- SONs, Self Organizing Networks methodologies.

Within the range of system development, LTE-Advanced and WiMAX 2 can use up to 8x8 MIMO and 128-QAM in downlink direction. Example performance: 100 MHz aggregated bandwidth, LTE-Advanced provides almost 3.3 Gbit peak download rates per sector of the base station under ideal conditions. Advanced network architectures combined with distributed and collaborative smart antenna technologies provide several years road map of commercial enhancements.

The 3GPP standards Release 12 added support for 256-QAM.

Timeframe and Introduction of Additional Features

Original standardization work for LTE-Advanced was done as part of 3GPP Release 10, which was frozen in April 2011. Trials were based on pre-release equipment. Major vendors support software upgrades to later versions and ongoing improvements.

In order to improve the quality of service for users in hotspots and on cell edges, heterogenous networks (HetNet) are formed of a mixture of macro-, pico- and femto base stations serving corresponding-size areas. Frozen in December 2012, 3GPP Release 11 concentrates on better support of HetNet. Coordinated Multi-Point operation (CoMP) is a key feature of Release 11 in order to support such network structures. Whereas users located at a cell edge in homogenous networks suffer from decreasing signal strength compounded by neighbor cell interference, CoMP is designed to enable use of a neighboring cell to also transmit the same signal as the serving cell, enhancing quality of service on the perimeter of a serving cell. In-device Co-existence (IDC) is another topic addressed in Release 11. IDC features are designed to ameliorate disturbances within the user equipment caused between LTE/LTE-A and the various other radio subsystems such as WiFi, Bluetooth, and the GPS receiver. Further enhancements for MIMO such as 4x4 configuration for the uplink were standardized.

The higher number of cells in HetNet results in user equipment changing the serving cell more frequently when in motion. The ongoing work on LTE-Advanced in Release 12, amongst other areas, concentrates on addressing issues that come about when users move through HetNet, such as frequent hand-overs between cells. It also included use of 256-QAM.

Voice Over LTE

Voice over Long Term Evolution (Voice over LTE or VoLTE) is the practice of packetizing voice over the Internet Protocol (VoIP) and transporting both the signaling and media components over a 4G LTE packet switched (PS) data path. This is in contrast to delivering voice using circuit switch (CS) methodologies, which requires 4G handsets to employ a secondary 3G radio and network operators to continue supporting this inefficient access infrastructure and licensed spectrum.

Originating in 2010, the GSM Association's Permanent Reference Document (PRD) IR.92 describes the application of the IP Multimedia Subsystem (IMS) for the transport of voice and short message service (SMS) traffic (SM-over-IP). IR.92 identifies a minimum subset of mandatory 3GPP (release 8) standard features that a mobile endpoints — or user equipment (UE) — are required to implement in order to guarantee interoperability in all-packet infrastructures. Along with defining basic IMS capabilities and the requirements of the Radio Access Network (RAN) and the Evolved Packet Core (EPC), this includes the 19 mandatory supplementary services which must be supported by the signaling core and the Telephony Application Server (TAS).

In addition, IR.92 defines the process by which calls fallback to circuit switched voice in the event that the UE roams out of a 4G LTE coverage area.

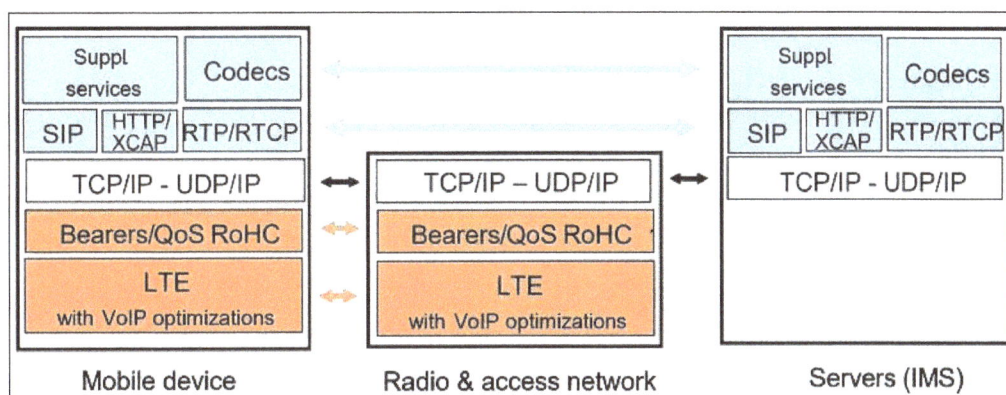

GSMA PRD IR.92 Depiction of UE and Network Protocol Stacks in IMS Profile for Voice.

With an IMS core as its foundation, VoLTE employs the Session Initiation Protocol (SIP) for registration, authentication addressing, call establishment and call termination. The Session Description Protocol (SDP) is employed for RTP media and bandwidth negotiation. As a conversational voice service, VoLTE demands the use of the LTE QoS Class Identifier (QCI) number 1, which provides a guaranteed bit rate (GBR) with sub 100ms packet delay and an error rate not exceeding 10-2. The SIP MESSAGE header is also employed to support SM-over-IP, with IR.92 referencing the applicable 3GPP specifications.

IR.92 and VoLTE is often used synonymously with IR.94 and ViLTE, which is the GSMA PRD describing the transport of conversational video over IMS.

References

- Guowang Miao; Jens Zander; Ki Won Sung; Ben Slimane (2016). Fundamentals of Mobile Data Networks. Cambridge University Press. ISBN 1107143217

- How-gsm-works: geeksforgeeks.org, Retrieved 18 April, 2019

- 3rd Generation Partnership Project (November 2014). "3GGP TS45.001: Technical Specification Group GSM/ EDGE Radio Access Network; Physical layer on the radio path; General description". 12.1.0. Retrieved 2015-12-05

- Frequency-division-multiple-access-fdma: networkencyclopedia.com, Retrieved 11 January, 2019

- Hujun Yin and Siavash Alamouti (August 2007). "OFDMA: A Broadband Wireless Access Technology". IEEE Sarnoff Symposium, 2006. IEEE: 1–4. Doi:10.1109/SARNOF.2006.4534773

Cellular Network Technology

- **Zero Generation**
- **First Generation**
- **Second Generation**
- **Third Generation**
- **Fourth Generation**
- **Fifth Generation**

There are a number of technologies which are used in cellular network to establish seamless communication. Some of these network technologies include zero generation, first generation, second-generation, third generation, fourth generation and fifth generation technology. This chapter closely examines these technologies of cellular network to provide an extensive understanding of the subject.

Zero Generation

OG (Zero Generation) is also known as Mobile Radio Telephone system. As this generation was invented prior to cellular system it was mentioned as pre cellular system. This system was analog in nature i.e. analog signals were used as carriers. Generally Mobile Radio Telephone system provides half duplex communications i.e. only one person will speak and other should hear. Mobile Radio Telephone system (Zero generation) consists of various technologies such as Advanced Mobile Telephone System (AMTS), Mobile Telephone System (MTS), MTD (Mobile telephony system D), OLT (Offentlig Landmobile Telefoni or Public Land Mobile Telephony), Push to Talk (PTT) and Improved Mobile Telephone Service (IMTS).These mobile telephones were placed in vehicles (truck, cars etc). The mobile telephone instrument had two main parts those were transceiver (transmitter – receiver) and head (instrument which had display and dial keys). Transceiver (transmitter – receiver) was fixed in the vehicle trunk; head was fixed near the driver seat and both head and transceiver were connected to each other with wire. The device (telephone) would connect to local telephone network only if it is in the range of 20 Kms. Each city had a central antenna tower with 25 channels. This means that mobile transceiver should have a powerful transmitter with a transmitting range of 50-70 Kms. Only few people were able to use this device as only 25 channels were available. Roaming facility was not supported in this generation of analog cellular Mobile Radio telephone system. Mobile Radio telephone system was a commercial service under public switched telephone network with unique telephone numbers. Zero generation had seen different variants of two way radio telephones. Large number of limitations in this generation led to the advent of new generation.

What is 0.5G Technology?

0.5G was the advance version of 0G (Zero Generation or Mobile Radio Telephone system). This 0.5G technology had introduced ARP (Autoradiopuhelin) as the first commercial public mobile phone network. This ARP network was launched in 1971 at Finland. ARP was operated on 8 Channels with a frequency of 150 MHz (147.9 – 154.875 MHz band) and its transmission power was in a range of 1 to 5 watts. ARP used half duplex system for transmission (voice signals can either be transmitted or received at a time) with manual switched system. This Network contains cells (Land area was divided into small sectors, each sector is known as cell, a cell is covered by a radio network with one transceiver) with the cell size of 30 km. As ARP did not support the handover, calls would get disconnected while moving from one cell to another. ARP provided 100% coverage which attracted many users towards it. ARP was successful and became very popular until the network became congested. The ARP mobile terminals were too large to be fixed in cars and were expensive too. These limitations led to invent of Autotel. Autotel are also known as PALM (Public Automated Land Mobile). Autotel is a radio telephone service which in terms of technology lies between MTS and IMTS. It used digital signals for messages like call stepup, channel assignment, ringing, etc only voice channel was analog. This system used existent high-power VHF channels instead of cellular system. It was developed in Canada and Columbia.

First Generation

The First generation of wireless telecommunication technology is known as 1G was introduced in 1980. The main difference between then existing systems and 1G was invent of cellular technology and hence it is also known as First generation of analog cellular telephone. In 1G or First generation of wireless telecommunication technology the network contains many cells (Land area was divided into small sectors, each sector is known as cell, a cell is covered by a radio network with one transceiver) and so same frequency can be reused many times which results in great spectrum usage and thus increased the system capacity i.e. large number of users could be accommodated easily.

Use of cellular system in 1G or First generation of wireless telecommunication technology resulted in great spectrum usage. The First generation of wireless telecommunication technology used analog transmission techniques which were basically used for transmitting voice signals. 1G or first generation of wireless telecommunication technology also consist of various standards among which most popular were Advance Mobile Phone Service (AMPS), Nordic Mobile Telephone (NMT), Total Access Communication System (TACS). All of the standards in 1G use frequency modulation techniques for voice signals and all the handover decisions were taken at the Base Stations (BS). The spectrum within cell was divided into number of channels and every call is allotted a dedicated pair of channels. Data transmission between the wire part of connection and PSTN (Packet Switched Telephone Network) was done using packet-switched network.

Different standards of 1G were used worldwide like:

In 1982 Advance Mobile Phone Service (AMPS) was employed in United States and later it was used in Canada, Central America, South America, Australia, Argentina, Brazil, Burma, Brunei, Bangladesh, China ,Cambodia, Georgia, Hong Kong, Indonesia, Malaysia, Kazakhstan, Mexico, Mongolia, Nauru, New Zealand, Pakistan, Guinea, Philippines, Russia, Singapore, South Korea, Sri lanka, Tajikistan, Taiwan, Thailand, Vietnam, Western Samoa.

Total Access Communication System (TACS)/Extended Total Access Communication System (ETACS) was employed in United Kingdom, United Arab Emirates, Kuwait, Macao, Bahrain, Malta, Singapore.

Nordic Mobile Telephone-450 (NMT-450) was employed in Austria, Belgium, Czech Republic, Denmark, Finland, France, Germany, Hungary, Poland, Russia, Spain, Sweden, Thailand, turkey and Ukraine.

Nordic Mobile Telephone-900 (NMT-900) was employed in Cyprus, Denmark, Finland, France, Greenland, Netherlands, Norway, Switzerland and Thailand.

C-NETZ (C-NETZ in German refers to C Network which was the first cellular wireless telephone network in Germany) was employed in Germany, Portugal and South Africa.

Radiocom2000 was employed in France.

Radio Telephone Mobile System (RTMS) was employed in Italy.

Nippon Telephone and Telegraph (NTT) was first employed in Japan and later NTACS (Narrowband Total Access Communications System) and JTACS (Japanese Total Access Communication System) were also employed.

Use of Analog signals for data (in this case voice) transmission led to many problems those are:

- Analog Signals does not allow advance encryption methods hence there is no security of data i.e. anybody could listen to the conversion easily by simple techniques. The user identification number could be stolen easily and which could be used to make any call and the user whose identification number was stolen had to pay the call charges.

- Analog signals can easily be affected by interference and the call quality decreases.

Second Generation

2G technology means second-generation wireless telephone technology. It is based on the technology known as the global system for the mobile communication or in short we can say GSM, This technology enabled various networks to offer services such as the text messages, the picture messages & MMS (multimedia messages).

2G Technology

GSM has enabled the users to make use of the short message services (SMS) to any mobile network at any time. SMS is a cheap & easy way to send a message to anyone, other than the voice call or conference. This technology is beneficial to both the network operators & ultimate users at the same time.

Second generation was launched in Finland in the year 1991, all phone conversations are digitally encrypted, and 2G networks are more efficient on the spectrum. They can allow far greater mobile phone penetration levels and 2G introduced data services for mobile, starting with SMS text messages.

2G technology is more efficient 2G network holds sufficient security for both the sender & the receiver. All text messages are digitally encrypted. This digital encryption allows for the transfer of data in such a way that only the intended receiver can receive & read it. These digital signals consume less battery power, so, it helps in saving the battery of mobiles.

2G technology.

2G Technologies (second generation technologies) are either time division multiple access (TDMA) or code division multiple access (CDMA). TDMA allows for the division of signal into time slots. CDMA allocates each user the special code to communicate over a multiplex physical channel.

Interim standard 95 is a first & the foremost CDMA cellular technology. It is most famous by its brand name known as cdmaOne. It makes use of the CDMA to transfer the voice signals & data signals from cellular phones to cell sites (cell sites is cellular network).

The digital systems were designed to emit less radio power from the handsets. This meant that the cells had to be smaller, so more cells had to be placed in the same amount of space. This was possible because the cell towers & related equipment had become less expensive.

2G Technology Advantages

2G technology is useful to both the users & the network operators at the same time. Digital systems are embraced by the consumers for many reasons. The lower powered radio signals require less battery power. The phones last longer between the charges & the batteries can be smaller.

2G technology offers improved privacy that was not possible with the earlier technologies. 2G phones are more private than 1G phones which have no protection whatsoever against eavesdropping. The digital cellular calls are harder to eavesdrop on by use of the radio scanners. While the security algorithms used have proved not to be as secure as initially advertised.

The digital calls tend to be free of static & background noise. The digital signals require very little battery power. The mobile phone connections in different countries of the world are based on digital signals and 2g technology.

The mobile batteries can last longer as the digital signals consume less battery power; so, they help the mobile batteries to last long. The digital coding reduces the noise in the line, thus improving the voice clarity and the digital signals are considered environment friendly.

From the other advantages of 2g technologies is that the lower power emissions have helped in dealing with health concerns. Nobody wants any unforeseen health concerns arising due to the use of any technology.

2G technologies introduce the digital data services such as SMS & email that has allowed the world to shrink & come closer. You cannot have two or more cloned handsets having the same phone number, under 2G technologies. This has helped in reducing any chances of fraud to a minimum.

The digital data service is used to assist the mobile network operators to introduce short message service over the cellular phones. The digital encryption has provided privacy & safety to the data & the voice calls, SMS is a cheap & easy way to communicate with anyone.

2G Technology Disadvantages

You can see many advantages of 2g technology. But using the 2G technology requires powerful digital signals to help the mobile phones work. But the digital signals could be weak if there is no network coverage in any specific area.

The weaker digital signal transmitted by the cellular phone cannot be sufficient to reach the cell tower in less populous areas. It causes a particular problem on 2G systems deployed on higher frequencies, but is mostly not a problem on 2G systems deployed on lower frequencies. The national regulations differ greatly among the countries which dictate where 2G can be deployed.

The digital signal has jagged decay curve, unlike the Analog that has a smooth decay curve. Under unfavorable conditions the digital will have occasional dropouts and may fail completely if the conditions worsen. As distance increases, the analog reception degrades gradually, but digital reception abruptly transitions from clear reception to no reception.

The digital will start to completely fail, by dropping calls or being unintelligible, while analog slowly gets worse, generally holding a call longer and allowing at least some of the audio transmitted to be understood.

Although the digital calls are free of static & background noise, the use of lossy compression by the codecs takes a toll and the range of sound that they transmit is reduced. You may hear less of the tonality of someone's voice talking on the digital cellphone.

The pulse nature of TDMA transmission used interferes with some electronics, such as certain audio amplifiers because the intellectual property is concerted among a few industry members. It can create the obstacles for new entrants and it limits the competition among the phone manufacturers.

2G is less compatible with the smartphones functions. Data transmission speed can be more than 4 million bits per sec in 3G technology but it is less than 50,000 bits per sec in 2G network. GSM offers a fixed maximum cell site range of 35 km which is imposed by the technical limitations. One could have a backup handset in case of damage or loss, a permanently installed handset in a car or remote workshop and so on. With digital systems, this is no longer possible, unless the two handsets are never turned on simultaneously. The downloading & uploading speeds available in 2G technologies are up to 236 Kbps. While in 3G technology the downloading and uploading speeds are up to 21 Mbps and 5.7 Mbps respectively.

Third Generation

3G, short for third generation, is the third generation of wireless mobile telecommunications technology. It is the upgrade for 2G and 2.5G GPRS networks, for faster data transfer speed. This is based on a set of standards used for mobile devices and mobile telecommunications use services and networks that comply with the International Mobile Telecommunications-2000 (IMT-2000) specifications by the International Telecommunication Union. 3G finds application in wireless voice telephony, mobile Internet access, fixed wireless Internet access, video calls and mobile TV.

3G telecommunication networks support services that provide an information transfer rate of at least 144 kbit/s. Later 3G releases, often denoted 3.5G and 3.75G, also provide mobile broadband access of several Mbit/s to smartphones and mobile modems in laptop computers. This ensures it can be applied to wireless voice telephony, mobile Internet access, fixed wireless Internet access, video calls and mobile TV technologies.

A new generation of cellular standards has appeared approximately every tenth year since 1G systems were introduced in 1979 and the early to mid-1980s. Each generation is characterized by new frequency bands, higher data rates and non−backward-compatible transmission technology. The first commercial 3G networks were introduced in 2000.

Several telecommunications companies market wireless mobile Internet services as *3G*, indicating that the advertised service is provided over a 3G wireless network. Services advertised as 3G are required to meet IMT-2000 technical standards, including standards for reliability and speed (data transfer rates). To meet the IMT-2000 standards, a system is required to provide peak data rates of at least 144 kbit/s. However, many services advertised as 3G provide higher speed than the minimum technical requirements for a 3G service. Recent 3G releases, often denoted 3.5G and 3.75G, also provide mobile broadband access of several Mbit/s to smartphones and mobile modems in laptop computers.

The following standards are typically branded 3G:

- The UMTS (Universal Mobile Telecommunications System) system, first offered in 2001, standardized by 3GPP, used primarily in Europe, Japan, China (however with a different radio interface) and other regions predominated by GSM (Global Systems for Mobile) 2G system infrastructure. The cell phones are typically UMTS and GSM hybrids. Several radio interfaces are offered, sharing the same infrastructure:

  ○ The original and most widespread radio interface is called W-CDMA (Wideband Code Division Multiple Access).

  ○ The TD-SCDMA radio interface was commercialized in 2009 and is only offered in China.

  ○ The latest UMTS release, HSPA+, can provide peak data rates up to 56 Mbit/s in the downlink in theory (28 Mbit/s in existing services) and 22 Mbit/s in the uplink.

- The CDMA2000 system, first offered in 2002, standardized by 3GPP2, used especially in North America and South Korea, sharing infrastructure with the IS-95 2G standard. The cell phones are typically CDMA2000 and IS-95 hybrids. The latest release EVDO Rev B offers peak rates of 14.7 Mbit/s downstream.

The above systems and radio interfaces are based on spread spectrum radio transmission technology. While the GSM EDGE standard ("2.9G"), DECT cordless phones and Mobile WiMAX standards formally also fulfill the IMT-2000 requirements and are approved as 3G standards by ITU, these are typically not branded 3G, and are based on completely different technologies.

The following common standards comply with the IMT2000/3G standard:

- EDGE, a revision by the 3GPP organization to the older 2G GSM based transmission methods, utilizing the same switching nodes, base station sites and frequencies as GPRS, but new base station and cellphone RF circuits. It is based on the three times as efficient 8PSK modulation scheme as supplement to the original GMSK modulation scheme. EDGE is still used extensively due to its ease of upgrade from existing 2G GSM infrastructure and cell-phones.

  ○ EDGE combined with the GPRS 2.5G technology is called EGPRS, and allows peak data rates in the order of 200 kbit/s, just as the original UMTS WCDMA versions, and thus formally fulfills the IMT2000 requirements on 3G systems. However, in practice EDGE is seldom marketed as a 3G system, but a 2.9G system. EDGE shows slightly better system spectral efficiency than the original UMTS and CDMA2000 systems, but it is difficult to reach much higher peak data rates due to the limited GSM spectral bandwidth of 200 kHz, and it is thus a dead end.

  ○ EDGE was also a mode in the IS-136 TDMA system, today ceased.

  ○ Evolved EDGE, the latest revision, has peaks of 1 Mbit/s downstream and 400 kbit/s upstream, but is not commercially used.

- The Universal Mobile Telecommunications System created and revised by the 3GPP. The family is a full revision from GSM in terms of encoding methods and hardware, although some GSM sites can be retrofitted to broadcast in the UMTS/W-CDMA format.

- W-CDMA is the most common deployment, commonly operated on the 2,100 MHz band. A few others use the 10, 900 and 1,900 MHz bands.

  - HSPA is an amalgamation of several upgrades to the original W-CDMA standard and offers speeds of 14.4 Mbit/s down and 5.76 Mbit/s up. HSPA is backward-compatible with and uses the same frequencies as W-CDMA.

  - HSPA+, a further revision and upgrade of HSPA, can provide theoretical peak data rates up to 168 Mbit/s in the downlink and 22 Mbit/s in the uplink, using a combination of air interface improvements as well as multi-carrier HSPA and MIMO. Technically though, MIMO and DC-HSPA can be used without the "+" enhancements of HSPA+.

- The CDMA2000 system, or IS-2000, including CDMA2000 1x and CDMA2000 High Rate Packet Data (or EVDO), standardized by 3GPP2 (*differing* from the 3GPP), evolving from the original IS-95 CDMA system, is used especially in North America, China, India, Pakistan, Japan, South Korea, Southeast Asia, Europe and Africa.

  - CDMA2000 1x Rev. E has an increased voice capacity (in excess of three times) compared to Rev. 0 EVDO Rev. B offers downstream peak rates of 14.7 Mbit/s while Rev. C enhanced existing and new terminal user experience.

While DECT cordless phones and Mobile WiMAX standards formally also fulfill the IMT-2000 requirements, they are not usually considered due to their rarity and unsuitability for usage with mobile phones.

Break-up of 3G Systems

The 3G (UMTS and CDMA2000) research and development projects started in 1992. In 1999, ITU approved five radio interfaces for IMT-2000 as a part of the ITU-R M.1457 Recommendation; WiMAX was added in 2007.

There are evolutionary standards (EDGE and CDMA) that are backward-compatible extensions to pre-existing 2G networks as well as revolutionary standards that require all-new network hardware and frequency allocations. The cell phones use UMTS in combination with 2G GSM standards and bandwidths, but *do not support EDGE*. The latter group is the UMTS family, which consists of standards developed for IMT-2000, as well as the independently developed standards DECT and WiMAX, which were included because they fit the IMT-2000 definition.

While EDGE fulfills the 3G specifications, most GSM/UMTS phones report EDGE ("2.75G") and UMTS ("3G") functionality.

Adoption

Japan was one of the first countries to adopt 3G, the reason being the process of 3G spectrum allocations, which in Japan was awarded without much upfront cost. Frequency spectrum was allocated in the US and Europe based on auctioning, thereby requiring a huge initial investment for any company wishing to provide 3G services. European companies collectively paid over 100 billion dollars in their spectrum auctions.

Nepal Telecom adopted 3G Service for the first time in southern Asia. However, its 3G was relatively slow to be adopted in Nepal. In some instances, 3G networks do not use the same radio frequencies as 2G so mobile operators must build entirely new networks and license entirely new frequencies, especially so to achieve high data transmission rates. Other countries' delays were due to the expenses of upgrading transmission hardware, especially for UMTS, whose deployment required the replacement of most broadcast towers. Due to these issues and difficulties with deployment, many carriers were not able to or delayed acquisition of these updated capabilities.

In December 2007, 190 3G networks were operating in 40 countries and 154 HSDPA networks were operating in 71 countries, according to the Global Mobile Suppliers Association (GSA). In Asia, Europe, Canada and the USA, telecommunication companies use W-CDMA technology with the support of around 100 terminal designs to operate 3G mobile networks.

Roll-out of 3G networks was delayed in some countries by the enormous costs of additional spectrum licensing fees. The license fees in some European countries were particularly high, bolstered by government auctions of a limited number of licenses and sealed bid auctions, and initial excitement over 3G's potential. This led to a telecoms crash that ran concurrently with similar crashes in the fibre-optic and dot.com fields.

The 3G standard is perhaps well known because of a massive expansion of the mobile communications market post-2G and advances of the consumer mobile phone. An especially notable development during this time is the smartphone (for example, the iPhone, and the Android family), combining the abilities of a PDA with a mobile phone, leading to widespread demand for mobile internet connectivity. 3G has also introduced the term "mobile broadband" because its speed and capability make it a viable alternative for internet browsing, and USB Modems connecting to 3G networks are becoming increasingly common.

Market Penetration

By June 2007, the 200 millionth 3G subscriber had been connected of which 10 million were in Nepal and 8.2 million in India. This 200 millionth is only 6.7% of the 3 billion mobile phone subscriptions worldwide. (When counting CDMA2000 1x RTT customers—max bitrate 72% of the 200kbit/s which defines 3G—the total size of the nearly-3G subscriber base was 475 million as of June 2007, which was 15.8% of all subscribers worldwide). In the countries where 3G was launched first – Japan and South Korea – 3G penetration is over 70%. In Europe the leading country for 3G penetration is Italy with a third of its subscribers migrated to 3G. Other leading countries for 3G use include Nepal, UK, Austria, Australia and Singapore at the 32% migration level.

According to ITU estimates, as of Q4 2012 there were 2096 million active mobile-broadband subscribers worldwide out of a total of 6835 million subscribers—this is just over 30%. About half the mobile-broadband subscriptions are for subscribers in developed nations, 934 million out of 1600 million total, well over 50%. Note however that there is a distinction between a phone with mobile-broadband connectivity and a smart phone with a large display and so on—although according to the ITU and informatandm.com the USA has 321 million mobile subscriptions, including 256 million that are 3G or 4G, which is both 80% of the subscriber base and 80% of the USA population, according to ComScore just a year earlier in Q4 2011 only about 42% of people surveyed in the

USA reported they owned a smart phone. In Japan, 3G penetration was similar at about 81%, but smart phone ownership was lower at about 17%. In China, there were 486.5 million 3G subscribers in June 2014, in a population of 1,385,566,537.

Features

Data Rates

ITU has not provided a clear definition of the data rate that users can expect from 3G equipment or providers. Thus users sold 3G service may not be able to point to a standard and say that the rates it specifies are not being met. While stating in commentary that "it is expected that IMT-2000 will provide higher transmission rates: a minimum data rate of 2 Mbit/s for stationary or walking users, and 348 kbit/s in a moving vehicle," the ITU does not actually clearly specify minimum required rates, nor required average rates, nor what modes of the interfaces qualify as 3G, so various data rates are sold as '3G' in the market.

In market implementation, 3G downlink data speeds defined by telecom service providers vary depending on the underlying technology deployed; up to 384kbit/s for WCDMA, up to 7.2Mbit/sec for HSPA and a theoretical maximum of 21.6 Mbit/s for HSPA+ (technically 3.5G, but usually clubbed under the tradename of 3G).

Compare data speeds with 3.5G and 4G.

Security

3G networks offer greater security than their 2G predecessors. By allowing the UE (User Equipment) to authenticate the network it is attaching to, the user can be sure the network is the intended one and not an impersonator. 3G networks use the KASUMI block cipher instead of the older A5/1 stream cipher. However, a number of serious weaknesses in the KASUMI cipher have been identified.

In addition to the 3G network infrastructure security, end-to-end security is offered when application frameworks such as IMS are accessed, although this is not strictly a 3G property.

Applications of 3G

The bandwidth and location information available to 3G devices gives rise to applications not previously available to mobile phone users.

Universal Mobile Telecommunications System

The Universal Mobile Telecommunications System (UMTS) is a third generation mobile cellular system for networks based on the GSM standard. Developed and maintained by the 3GPP (3rd Generation Partnership Project), UMTS is a component of the International Telecommunications Union IMT-2000 standard set and compares with the CDMA2000 standard set for networks based on the competing cdmaOne technology. UMTS uses wideband code division multiple access (W-CDMA) radio access technology to offer greater spectral efficiency and bandwidth to mobile network operators.

UMTS specifies a complete network system, which includes the radio access network (UMTS Terrestrial Radio Access Network, or UTRAN), the core network (Mobile Application Part, or MAP) and the authentication of users via SIM (subscriber identity module) cards.

The technology described in UMTS is sometimes also referred to as Freedom of Mobile Multimedia Access (FOMA) or 3GSM.

Unlike EDGE (IMT Single-Carrier, based on GSM) and CDMA2000 (IMT Multi-Carrier), UMTS requires new base stations and new frequency allocations.

Features

UMTS supports maximum theoretical data transfer rates of 42 Mbit/s when Evolved HSPA (HSPA+) is implemented in the network. Users in deployed networks can expect a transfer rate of up to 384 kbit/s for Release '99 (R99) handsets (the original UMTS release), and 7.2 Mbit/s for High-Speed Downlink Packet Access (HSDPA) handsets in the downlink connection. These speeds are significantly faster than the 9.6 kbit/s of a single GSM error-corrected circuit switched data channel, multiple 9.6 kbit/s channels in High-Speed Circuit-Switched Data (HSCSD) and 14.4 kbit/s for CDMAOne channels.

Since 2006, UMTS networks in many countries have been or are in the process of being upgraded with High-Speed Downlink Packet Access (HSDPA), sometimes known as 3.5G. Currently, HSDPA enables downlink transfer speeds of up to 21 Mbit/s. Work is also progressing on improving the uplink transfer speed with the High-Speed Uplink Packet Access (HSUPA). Longer term, the 3GPP Long Term Evolution (LTE) project plans to move UMTS to 4G speeds of 100 Mbit/s down and 50 Mbit/s up, using a next generation air interface technology based upon orthogonal frequency-division multiplexing.

The first national consumer UMTS networks launched in 2002 with a heavy emphasis on telco-provided mobile applications such as mobile TV and video calling. The high data speeds of UMTS are now most often utilised for Internet access: experience in Japan and elsewhere has shown that user demand for video calls is not high, and telco-provided audio/video content has declined in popularity in favour of high-speed access to the World Wide Web—either directly on a handset or connected to a computer via Wi-Fi, Bluetooth or USB.

Air Interfaces

UMTS combines three different terrestrial air interfaces, GSM's Mobile Application Part (MAP) core, and the GSM family of speech codecs.

The air interfaces are called UMTS Terrestrial Radio Access (UTRA). All air interface options are part of ITU's IMT-2000. In the currently most popular variant for cellular mobile telephones, W-CDMA (IMT Direct Spread) is used. It is also called "Uu interface", as it links User Equipment to the UMTS Terrestrial Radio Access Network.

The terms W-CDMA, TD-CDMA and TD-SCDMA are misleading. While they suggest covering just a channel access method (namely a variant of CDMA), they are actually the common names for the whole air interface standards.

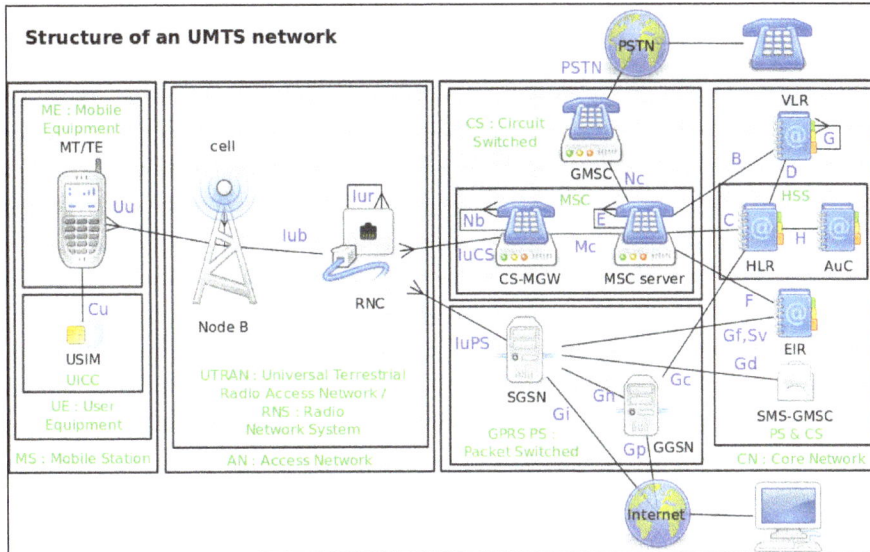

UMTS network architecture.

W-CDMA (UTRA-FDD)

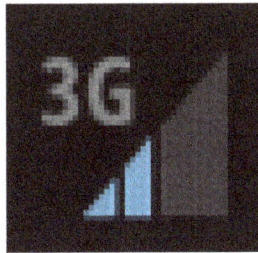

3G sign shown in notification bar on an Android powered smartphone.

W-CDMA or WCDMA (Wideband Code Division Multiple Access), along with UMTS-FDD, UTRA-FDD, or IMT-2000 CDMA Direct Spread is an air interface standard found in 3G mobile telecommunications networks. It supports conventional cellular voice, text and MMS services, but can also carry data at high speeds, allowing mobile operators to deliver higher bandwidth applications including streaming and broadband Internet access.

W-CDMA uses the DS-CDMA channel access method with a pair of 5 MHz wide channels. In contrast, the competing CDMA2000 system uses one or more available 1.25 MHz channels for each direction of communication. W-CDMA systems are widely criticized for their large spectrum usage, which delayed deployment in countries that acted relatively slowly in allocating new frequencies specifically for 3G services (such as the United States).

The specific frequency bands originally defined by the UMTS standard are 1885–2025 MHz for the mobile-to-base (uplink) and 2110–2200 MHz for the base-to-mobile. In the US, 1710–1755 MHz and 2110–2155 MHz are used instead, as the 1900 MHz band was already used. While UMTS2100 is the most widely deployed UMTS band, some countries' UMTS operators use the 850 MHz (900 MHz in Europe) and/or 1900 MHz bands (independently, meaning uplink and downlink are within the same band), notably in the US by AT&T Mobility, New Zealand by Telecom New Zealand on the XT Mobile Network and in Australia by Telstra on the Next G network. Some carriers such as T-Mobile use band numbers to identify the UMTS frequencies. For example, Band I (2100 MHz), Band IV (1700/2100 MHz), and Band V (850 MHz).

UMTS base station on the roof of a building.

UMTS-FDD is an acronym for Universal Mobile Telecommunications System (UMTS) - frequency-division duplexing (FDD) and a 3GPP standardized version of UMTS networks that makes use of frequency-division duplexing for duplexing over an UMTS Terrestrial Radio Access (UTRA) air interface.

W-CDMA is the basis of Japan's NTT DoCoMo's FOMA service and the most-commonly used member of the Universal Mobile Telecommunications System (UMTS) family and sometimes used as a synonym for UMTS. It uses the DS-CDMA channel access method and the FDD duplexing method to achieve higher speeds and support more users compared to most previously used time division multiple access (TDMA) and time division duplex (TDD) schemes.

While not an evolutionary upgrade on the airside, it uses the same core network as the 2G GSM networks deployed worldwide, allowing dual mode mobile operation along with GSM/EDGE; a feature it shares with other members of the UMTS family.

Development

In the late 1990s, W-CDMA was developed by NTT DoCoMo as the air interface for their 3G network FOMA. Later NTT DoCoMo submitted the specification to the International Telecommunication Union (ITU) as a candidate for the international 3G standard known as IMT-2000. The ITU eventually accepted W-CDMA as part of the IMT-2000 family of 3G standards, as an alternative to CDMA2000, EDGE, and the short range DECT system. Later, W-CDMA was selected as an air interface for UMTS.

As NTT DoCoMo did not wait for the finalisation of the 3G Release 99 specification, their network was initially incompatible with UMTS. However, this has been resolved by NTT DoCoMo updating their network.

Code Division Multiple Access communication networks have been developed by a number of companies over the years, but development of cell-phone networks based on CDMA (prior to W-CDMA) was dominated by Qualcomm, the first company to succeed in developing a practical and cost-effective CDMA implementation for consumer cell phones and its early IS-95 air interface standard has evolved into the current CDMA2000 (IS-856/IS-2000) standard. Qualcomm created an experimental wideband CDMA system called CDMA2000 3x which unified the W-CDMA (3GPP) and CDMA2000 (3GPP2) network technologies into a single design for a worldwide standard air interface. Compatibility with CDMA2000 would have beneficially enabled roaming on existing networks beyond Japan, since Qualcomm CDMA2000 networks are widely deployed, especially in the Americas, with coverage in 58 countries as of 2006. However, divergent requirements resulted in the W-CDMA standard being retained and deployed globally. W-CDMA has then become the dominant technology with 457 commercial networks in 178 countries as of April 2012. Several CDMA2000 operators have even converted their networks to W-CDMA for international roaming compatibility and smooth upgrade path to LTE.

Despite incompatibility with existing air-interface standards, late introduction and the high upgrade cost of deploying an all-new transmitter technology, W-CDMA has become the dominant standard.

Rationale for W-CDMA

W-CDMA transmits on a pair of 5 MHz-wide radio channels, while CDMA2000 transmits on one or several pairs of 1.25 MHz radio channels. Though W-CDMA does use a direct sequence CDMA transmission technique like CDMA2000, W-CDMA is not simply a wideband version of CDMA2000. The W-CDMA system is a new design by NTT DoCoMo, and it differs in many aspects from CDMA2000. From an engineering point of view, W-CDMA provides a different balance of trade-offs between cost, capacity, performance, and density; it also promises to achieve a benefit of reduced cost for video phone handsets. W-CDMA may also be better suited for deployment in the very dense cities of Europe and Asia. However, hurdles remain, and cross-licensing of patents between Qualcomm and W-CDMA vendors has not eliminated possible patent issues due to the features of W-CDMA which remain covered by Qualcomm patents.

W-CDMA has been developed into a complete set of specifications, a detailed protocol that defines how a mobile phone communicates with the tower, how signals are modulated, how datagrams are structured, and system interfaces are specified allowing free competition on technology elements.

Deployment

The world's first commercial W-CDMA service, FOMA, was launched by NTT DoCoMo in Japan in 2001. Elsewhere, W-CDMA deployments are usually marketed under the UMTS brand. W-CDMA has also been adapted for use in satellite communications on the U.S. Mobile User Objective System using geosynchronous satellites in place of cell towers.

J-Phone Japan (once Vodafone and now SoftBank Mobile) soon followed by launching their own W-CDMA based service, originally branded "Vodafone Global Standard" and claiming UMTS compatibility. The name of the service was changed to "Vodafone 3G" (now "SoftBank 3G") in December 2004.

Beginning in 2003, Hutchison Whampoa gradually launched their upstart UMTS networks. Most countries have, since the ITU approved of the 3G mobile service, either "auctioned" the radio frequencies to the company willing to pay the most, or conducted a "beauty contest"—asking the various companies to present what they intend to commit to if awarded the licences. This strategy has been criticised for aiming to drain the cash of operators to the brink of bankruptcy in order to honour their bids or proposals. Most of them have a time constraint for the rollout of the service—where certain "coverage" must be achieved within a given date or the licence will be revoked.

Vodafone launched several UMTS networks in Europe in February 2004. MobileOne of Singapore commercially launched its 3G (W-CDMA) services in February 2005. New Zealand in August 2005 and Australia in October 2005.

AT&T Wireless (now a part of Cingular Wireless) has deployed UMTS in several cities. Though advancements in its network deployment have been delayed due to the merger with Cingular, Cingular began offering HSDPA service in December 2005.

Rogers in Canada March 2007 has launched HSDPA in the Toronto Golden Horseshoe district on W-CDMA at 850/1900 MHz and plan the launch the service commercial in the top 25 cities October, 2007.

TeliaSonera opened W-CDMA service in Finland October 13, 2004 with speeds up to 384 kbit/s. Availability only in main cities. Pricing is approx. €2/MB. K Telecom and KTF, two largest mobile phone service providers in South Korea, have each started offering W-CDMA service in December 2003. Due to poor coverage and lack of choice in handhelds, the W-CDMA service has barely made a dent in the Korean market which was dominated by CDMA2000. By October 2006 both companies are covering more than 90 cities while SK Telecom has announced that it will provide nationwide coverage for its WCDMA network in order for it to offer SBSM (Single Band Single Mode) handsets by the first half of 2007. KT Freecel will thus cut funding to its CDMA2000 network development to the minimum.

In Norway, Telenor introduced W-CDMA in major cities by the end of 2004, while their competitor, NetCom, followed suit a few months later. Both operators have 98% national coverage on EDGE, but Telenor has parallel WLAN roaming networks on GSM, where the UMTS service is competing with this. For this reason Telenor is dropping support of their WLAN service in Austria (2006).

Maxis Communications and Celcom, two mobile phone service providers in Malaysia, started offering W-CDMA services in 2005. In Sweden, Telia introduced W-CDMA in March 2004.

UTRA-TDD

UMTS-TDD, an acronym for Universal Mobile Telecommunications System (UMTS) - time-division duplexing (TDD), is a 3GPP standardized version of UMTS networks that use UTRA-TDD. UTRA-TDD is a UTRA that uses time-division duplexing for duplexing. While a full implementation of UMTS, it is mainly used to provide Internet access in circumstances similar to those where WiMAX might be used. UMTS-TDD is not directly compatible with UMTS-FDD: a device designed to use one standard cannot, unless specifically designed to, work on the other, because of the difference in air interface technologies and frequencies used. It is more formally as IMT-2000 CDMA-TDD or IMT 2000 Time-Division (IMT-TD).

The two UMTS air interfaces (UTRAs) for UMTS-TDD are TD-CDMA and TD-SCDMA. Both air interfaces use a combination of two channel access methods, code division multiple access (CDMA) and time division multiple access (TDMA): the frequency band is divided into time slots (TDMA), which are further divided into channels using CDMA spreading codes. These air interfaces are classified as TDD, because time slots can be allocated to either uplink or downlink traffic.

TD-CDMA (UTRA-TDD 3.84 Mcps High Chip Rate (HCR))

TD-CDMA, an acronym for Time-division-Code division multiple access, is a channel access method based on using spread spectrum multiple access (CDMA) across multiple time slots (TDMA). TD-CDMA is the channel access method for UTRA-TDD HCR, which is an acronym for UMTS Terrestrial Radio Access-Time Division Duplex High Chip Rate.

UMTS-TDD's air interfaces that use the TD-CDMA channel access technique are standardized as UTRA-TDD HCR, which uses increments of 5 MHz of spectrum, each slice divided into 10 ms frames containing fifteen time slots (1500 per second). The time slots (TS) are allocated in fixed percentage for downlink and uplink. TD-CDMA is used to multiplex streams from or to multiple transceivers. Unlike W-CDMA, it does not need separate frequency bands for up- and downstream, allowing deployment in tight frequency bands.

TD-CDMA is a part of IMT-2000, defined as IMT-TD Time-Division (IMT CDMA TDD), and is one of the three UMTS air interfaces (UTRAs), as standardized by the 3GPP in UTRA-TDD HCR. UTRA-TDD HCR is closely related to W-CDMA, and provides the same types of channels where possible. UMTS's HSDPA/HSUPA enhancements are also implemented under TD-CDMA.

In the United States, the technology has been used for public safety and government use in the New York City and a few other areas. In Japan, IPMobile planned to provide TD-CDMA service in year 2006, but it was delayed, changed to TD-SCDMA, and bankrupt before the service officially started.

TD-SCDMA (UTRA-TDD 1.28 Mcps Low Chip Rate (LCR))

Time Division Synchronous Code Division Multiple Access (TD-SCDMA) or UTRA TDD 1.28 mcps low chip rate (UTRA-TDD LCR) is an air interface found in UMTS mobile telecommunications networks in China as an alternative to W-CDMA.

TD-SCDMA uses the TDMA channel access method combined with an adaptive synchronous CDMA component on 1.6 MHz slices of spectrum, allowing deployment in even tighter frequency bands than TD-CDMA. It is standardized by the 3GPP and also referred to as "UTRA-TDD LCR". However, the main incentive for development of this Chinese-developed standard was avoiding or reducing the license fees that have to be paid to non-Chinese patent owners. Unlike the other air interfaces, TD-SCDMA was not part of UMTS from the beginning but has been added in Release 4 of the specification.

Like TD-CDMA, TD-SCDMA is known as IMT CDMA TDD within IMT-2000. The term "TD-SCD-MA" is misleading. While it suggests covering only a channel access method, it is actually the common name for the whole air interface specification.

TD-SCDMA / UMTS-TDD (LCR) networks are incompatible with W-CDMA / UMTS-FDD and TD-CDMA / UMTS-TDD (HCR) networks.

Objectives

TD-SCDMA was developed in the People's Republic of China by the Chinese Academy of Telecommunications Technology (CATT), Datang Telecom, and Siemens AG in an attempt to avoid dependence on Western technology. This is likely primarily for practical reasons, since other 3G formats require the payment of patent fees to a large number of Western patent holders.

TD-SCDMA proponents also claim it is better suited for densely populated areas. Further, it is supposed to cover all usage scenarios, whereas W-CDMA is optimised for symmetric traffic and macro cells, while TD-CDMA is best used in low mobility scenarios within micro or pico cells.

TD-SCDMA is based on spread spectrum technology which makes it unlikely that it will be able to completely escape the payment of license fees to western patent holders. The launch of a national TD-SCDMA network was initially projected by 2005 but only reached large scale commercial trials with 60,000 users across eight cities in 2008.

On January 7, 2009, China granted a TD-SCDMA 3G licence to China Mobile. On September 21, 2009, China Mobile officially announced that it had 1,327,000 TD-SCDMA subscribers as of the end of August, 2009.

While TD is primarily a China-only system, it may well be exported to developing countries. It is likely to be replaced with a newer TD-LTE system over the next 5 years.

Technical Highlights

TD-SCDMA uses TDD, in contrast to the FDD scheme used by W-CDMA. By dynamically adjusting the number of timeslots used for downlink and uplink, the system can more easily accommodate asymmetric traffic with different data rate requirements on downlink and uplink than FDD schemes. Since it does not require paired spectrum for downlink and uplink, spectrum allocation flexibility is also increased. Using the same carrier frequency for uplink and downlink also means that the channel condition is the same on both directions, and the base station can deduce the downlink channel information from uplink channel estimates, which is helpful to the application of beamforming techniques.

TD-SCDMA also uses TDMA in addition to the CDMA used in WCDMA. This reduces the number of users in each timeslot, which reduces the implementation complexity of multiuser detection and beamforming schemes, but the non-continuous transmission also reduces coverage (because of the higher peak power needed), mobility (because of lower power control frequency) and complicates radio resource management algorithms.

The "S" in TD-SCDMA stands for "synchronous", which means that uplink signals are synchronized at the base station receiver, achieved by continuous timing adjustments. This reduces the interference between users of the same timeslot using different codes by improving the orthogonality between the codes, therefore increasing system capacity, at the cost of some hardware complexity in achieving uplink synchronization.

Unlicensed UMTS-TDD

In Europe, CEPT allocated the 2010-2020 MHz range for a variant of UMTS-TDD designed for unlicensed, self-provided use. Some telecom groups and jurisdictions have proposed withdrawing this service in favour of licensed UMTS-TDD, due to lack of demand, and lack of development of a UMTS TDD air interface technology suitable for deployment in this band.

Comparison with UMTS-FDD

Ordinary UMTS uses UTRA-FDD as an air interface and is known as UMTS-FDD. UMTS-FDD uses W-CDMA for multiple access and frequency division for duplexing, meaning that the up-link and down-link transmit on different frequencies. UMTS is usually transmitted on frequencies assigned for 1G, 2G, or 3G mobile telephone service in the countries of operation.

UMTS-TDD uses time division duplexing, allowing the up-link and down-link to share the same spectrum. This allows the operator to more flexibly divide the usage of available spectrum according to traffic patterns. For ordinary phone service, you would expect the up-link and down-link to carry approximately equal amounts of data (because every phone call needs a voice transmission in either direction), but Internet-oriented traffic is more frequently one-way. For example, when browsing a website, the user will send commands, which are short, to the server, but the server will send whole files, that are generally larger than those commands, in response.

UMTS-TDD tends to be allocated frequency intended for mobile/wireless Internet services rather than used on existing cellular frequencies. This is, in part, because TDD duplexing is not normally allowed on cellular, PCS/PCN, and 3G frequencies. TDD technologies open up the usage of leftover unpaired spectrum.

Europe-wide, several bands are provided either specifically for UMTS-TDD or for similar technologies. These are 1900 MHz and 1920 MHz and between 2010 MHz and 2025 MHz. In several countries the 2500-2690 MHz band (also known as MMDS in the USA) has been used for UMTS-TDD deployments. Additionally, spectrum around the 3.5 GHz range has been allocated in some countries, notably Britain, in a technology-neutral environment. In the Czech Republic UTMS-TDD is also used in a frequency range around 872 MHz.

Deployment

UMTS-TDD has been deployed for public and/or private networks in at least nineteen countries around the world, with live systems in, amongst other countries, Australia, Czech Republic, France, Germany, Japan, New Zealand, Botswana, South Africa, the UK, and the USA.

Deployments in the US thus far have been limited. It has been selected for a public safety support network used by emergency responders in New York, but outside of some experimental systems, notably one from Nextel, thus far the WiMAX standard appears to have gained greater traction as a general mobile Internet access system.

Competing Standards

A variety of Internet-access systems exist which provide broadband speed access to the net. These include WiMAX and HIPERMAN. UMTS-TDD has the advantages of being able to use an operator's existing UMTS/GSM infrastructure, should it have one, and that it includes UMTS modes optimized for circuit switching should, for example, the operator want to offer telephone service. UMTS-TDD's performance is also more consistent. However, UMTS-TDD deployers often have regulatory problems with taking advantage of some of the services UMTS compatibility provides. For example, UMTS-TDD spectrum in the UK cannot be used to provide telephone service, though the regulator OFCOM is discussing the possibility of allowing it at some point in the future. Few operators considering UMTS-TDD have existing UMTS/GSM infrastructure.

Additionally, the WiMAX and HIPERMAN systems provide significantly larger bandwidths when the mobile station is in close proximity to the tower.

Like most mobile Internet access systems, many users who might otherwise choose UMTS-TDD will find their needs covered by the ad hoc collection of unconnected Wi-Fi access points at many restaurants and transportation hubs, and/or by Internet access already provided by their mobile phone operator. By comparison, UMTS-TDD (and systems like WiMAX) offers mobile, and more consistent, access than the former and generally faster access than the latter.

Radio Access Network

UMTS also specifies the Universal Terrestrial Radio Access Network (UTRAN), which is composed of multiple base stations, possibly using different terrestrial air interface standards and frequency bands.

UMTS and GSM/EDGE can share a Core Network (CN), making UTRAN an alternative radio access network to GERAN (GSM/EDGE RAN), and allowing (mostly) transparent switching between the RANs according to available coverage and service needs. Because of that, UMTS's and GSM/EDGE's radio access networks are sometimes collectively referred to as UTRAN/GERAN.

UMTS networks are often combined with GSM/EDGE, the latter of which is also a part of IMT-2000.

The UE (User Equipment) interface of the RAN (Radio Access Network) primarily consists of RRC (Radio Resource Control), PDCP (Packet Data Convergence Protocol), RLC (Radio Link Control) and MAC (Media Access Control) protocols. RRC protocol handles connection establishment, measurements, radio bearer services, security and handover decisions. RLC protocol primarily divides into three Modes—Transparent Mode (TM), Unacknowledge Mode (UM), Acknowledge Mode (AM). The functionality of AM entity resembles TCP operation whereas UM operation resembles UDP operation. In TM mode, data will be sent to lower layers without adding any header to SDU of higher layers. MAC handles the scheduling of data on air interface depending on higher layer (RRC) configured parameters.

The set of properties related to data transmission is called Radio Bearer (RB). This set of properties decides the maximum allowed data in a TTI (Transmission Time Interval). RB includes RLC information and RB mapping. RB mapping decides the mapping between RB<->logical channel<->transport channel. Signaling messages are sent on Signaling Radio Bearers (SRBs) and data packets (either CS or PS) are sent on data RBs. RRC and NAS messages go on SRBs.

Security includes two procedures: integrity and ciphering. Integrity validates the resource of messages and also makes sure that no one (third/unknown party) on the radio interface has modified the messages. Ciphering ensures that no one listens to your data on the air interface. Both integrity and ciphering are applied for SRBs whereas only ciphering is applied for data RBs.

Core Network

With Mobile Application Part, UMTS uses the same core network standard as GSM/EDGE. This allows a simple migration for existing GSM operators. However, the migration path to UMTS is still costly: while much of the core infrastructure is shared with GSM, the cost of obtaining new spectrum licenses and overlaying UMTS at existing towers is high.

The CN can be connected to various backbone networks, such as the Internet or an Integrated Services Digital Network (ISDN) telephone network. UMTS (and GERAN) include the three lowest layers of OSI model. The network layer (OSI 3) includes the Radio Resource Management protocol (RRM) that manages the bearer channels between the mobile terminals and the fixed network, including the handovers.

Frequency Bands and Channel Bandwidths

UARFCN

A UARFCN (abbreviation for UTRA Absolute Radio Frequency Channel Number, where UTRA stands for UMTS Terrestrial Radio Access) is used to identify a frequency in the UMTS frequency bands.

Typically channel number is derived from the frequency in MHz through the formula Channel Number = Frequency * 5. However, this is only able to represent channels that are centered on a multiple of 200 kHz, which do not align with licensing in North America. 3GPP added several special values for the common North American channels.

Spectrum Allocation

Over 130 licenses have already been awarded to operator's worldwide, specifying W-CDMA radio access technology that builds on GSM. In Europe, the license process occurred at the tail end of the technology bubble, and the auction mechanisms for allocation set up in some countries resulted in some extremely high prices being paid for the original 2100 MHz licenses, notably in the UK and Germany. In Germany, bidders paid a total €50.8 billion for six licenses, two of which were subsequently abandoned and written off by their purchasers (Mobilcom and the Sonera/Telefonica consortium). It has been suggested that these huge license fees have the character of a very large tax paid on future income expected many years down the road. In any event, the high prices paid put some European telecom operators close to bankruptcy (most notably KPN). Over the last few years some operators have written off some or all of the license costs. Between 2007 and 2009, all three Finnish carriers began to use 900 MHz UMTS in a shared arrangement with its surrounding 2G GSM base stations for rural area coverage, a trend that is expected to expand over Europe in the next 1–3 years.

The 2100 MHz band (downlink around 2100 MHz and uplink around 1900 MHz) allocated for UMTS in Europe and most of Asia is already used in North America. The 1900 MHz range is used for 2G (PCS) services, and 2100 MHz range is used for satellite communications. Regulators have, however, freed up some of the 2100 MHz range for 3G services, together with a different range around 1700 MHz for the uplink.

AT&T Wireless launched UMTS services in the United States by the end of 2004 strictly using the existing 1900 MHz spectrum allocated for 2G PCS services. Cingular acquired AT&T Wireless in 2004 and has since then launched UMTS in select US cities. Cingular renamed itself AT&T Mobility and rolled out some cities with a UMTS network at 850 MHz to enhance its existing UMTS network at 1900 MHz and now offers subscribers a number of dual-band UMTS 850/1900 phones.

T-Mobile's rollout of UMTS in the US was originally focused on the 1700 MHz band. However, T-Mobile has been moving users from 1700 MHz to 1900 MHz (PCS) in order to reallocate the spectrum to 4G LTE services.

In Canada, UMTS coverage is being provided on the 850 MHz and 1900 MHz bands on the Rogers and Bell-Telus networks. Bell and Telus share the network. Recently, new providers Wind Mobile, Mobilicity and Videotron have begun operations in the 1700 MHz band.

In 2008, Australian telco Telstra replaced its existing CDMA network with a national UMTS-based 3G network, branded as NextG, operating in the 850 MHz band. Telstra currently provides UMTS service on this network, and also on the 2100 MHz UMTS network, through a co-ownership of the owning and administrating company 3GIS. This company is also co-owned by Hutchison 3G Australia, and this is the primary network used by their customers. Optus is currently rolling out a 3G network operating on the 2100 MHz band in cities and most large towns, and the 900 MHz band in regional areas. Vodafone is also building a 3G network using the 900 MHz band.

Carriers in South America are now also rolling out 850 MHz networks.

Interoperability and Global Roaming

UMTS phones (and data cards) are highly portable—they have been designed to roam easily onto other UMTS networks (if the providers have roaming agreements in place). In addition, almost all UMTS phones are UMTS/GSM dual-mode devices, so if a UMTS phone travels outside of UMTS coverage during a call the call may be transparently handed off to available GSM coverage. Roaming charges are usually significantly higher than regular usage charges.

Most UMTS licensees consider ubiquitous, transparent global roaming an important issue. To enable a high degree of interoperability, UMTS phones usually support several different frequencies in addition to their GSM fallback. Different countries support different UMTS frequency bands – Europe initially used 2100 MHz while the most carriers in the USA use 850 MHz and 1900 MHz. T-Mobile has launched a network in the US operating at 1700 MHz (uplink) /2100 MHz (downlink), and these bands also have been adopted elsewhere in the US and in Canada and Latin America. A UMTS phone and network must support a common frequency to work together. Because of the frequencies used, early models of UMTS phones designated for the United States will likely not be operable elsewhere and vice versa. There are now 11 different frequency combinations used around the world—including frequencies formerly used solely for 2G services.

UMTS phones can use a Universal Subscriber Identity Module, USIM (based on GSM's SIM card) and also work (including UMTS services) with GSM SIM cards. This is a global standard of identification, and enables a network to identify and authenticate the (U)SIM in the phone. Roaming agreements between networks allow for calls to a customer to be redirected to them while roaming and determine the services (and prices) available to the user. In addition to user subscriber information and authentication information, the (U)SIM provides storage space for phone book contact. Handsets can store their data on their own memory or on the (U)SIM card (which is usually more limited in its phone book contact information). A (U)SIM can be moved to another UMTS or GSM phone, and the phone will take on the user details of the (U)SIM, meaning it is the (U)SIM (not the phone) which determines the phone number of the phone and the billing for calls made from the phone.

Japan was the first country to adopt 3G technologies, and since they had not used GSM previously they had no need to build GSM compatibility into their handsets and their 3G handsets were smaller than those available elsewhere. In 2002, NTT DoCoMo's FOMA 3G network was the first commercial UMTS network—using a pre-release specification, it was initially incompatible with the UMTS standard at the radio level but used standard USIM cards, meaning USIM card based roaming was possible (transferring the USIM card into a UMTS or GSM phone when travelling). Both NTT DoCoMo and SoftBank Mobile (which launched 3G in December 2002) now use standard UMTS.

Handsets and Modems

The Nokia 6650, an early (2003) UMTS handset.

All of the major 2G phone manufacturers (that are still in business) are now manufacturers of 3G phones. The early 3G handsets and modems were specific to the frequencies required in their country, which meant they could only roam to other countries on the same 3G frequency (though they can fall back to the older GSM standard). Canada and USA have a common share of frequencies, as do most European countries.

Using a cellular router, PCMCIA or USB card, customers are able to access 3G broadband services, regardless of their choice of computer (such as a tablet PC or a PDA). Some software installs itself from the modem, so that in some cases absolutely no knowledge of technology is required to get online in moments. Using a phone that supports 3G and Bluetooth 2.0, multiple Bluetooth-capable laptops can be connected to the Internet. Some smartphones can also act as a mobile WLAN access point.

There are very few 3G phones or modems available supporting all 3G frequencies (UMTS850/900/1700/1900/2100 MHz). Nokia has recently released a range of phones that have Pentaband 3G coverage, including the N8 and E7. Many other phones are offering more than one band which still enables extensive roaming. For example, Apple's iPhone 4 contains a quadband chipset operating on 850/900/1900/2100 MHz, allowing usage in the majority of countries where UMTS-FDD is deployed.

Other Competing Standards

The main competitor to UMTS is CDMA2000 (IMT-MC), which is developed by the 3GPP2. Unlike UMTS, CDMA2000 is an evolutionary upgrade to an existing 2G standard, cdmaOne, and is able to operate within the same frequency allocations. This and CDMA2000's narrower bandwidth requirements make it easier to deploy in existing spectra. In some, but not all, cases, existing GSM operators only have enough spectrums to implement either UMTS or GSM, not both. For example, in the US D, E, and F PCS spectrum blocks, the amount of spectrum available is 5 MHz in each direction. A standard UMTS system would saturate that spectrum. Where CDMA2000 is deployed, it usually co-exists with UMTS. In many markets however, the co-existence issue is of little relevance, as legislative hurdles exist to co-deploying two standards in the same licensed slice of spectrum.

Another competitor to UMTS is EDGE (IMT-SC), which is an evolutionary upgrade to the 2G GSM system, leveraging existing GSM spectrums. It is also much easier, quicker, and considerably cheaper for wireless carriers to "bolt-on" EDGE functionality by upgrading their existing GSM transmission hardware to support EDGE rather than having to install almost all brand-new equipment to deliver UMTS. However, being developed by 3GPP just as UMTS, EDGE is not a true competitor. Instead, it is used as a temporary solution preceding UMTS roll-out or as a complement for rural areas. This is facilitated by the fact that GSM/EDGE and UMTS specification are jointly developed and rely on the same core network, allowing dual-mode operation including vertical handovers.

China's TD-SCDMA standard is often seen as a competitor, too. TD-SCDMA has been added to UMTS' Release 4 as UTRA-TDD 1.28 Mcps Low Chip Rate (UTRA-TDD LCR). Unlike TD-CDMA (UTRA-TDD 3.84 Mcps High Chip Rate, UTRA-TDD HCR) which complements W-CDMA (UTRA-FDD), it is suitable for both micro and macro cells. However, the lack of vendors' support is preventing it from being a real competitor.

While DECT is technically capable of competing with UMTS and other cellular networks in densely populated, urban areas, it has only been deployed for domestic cordless phones and private in-house networks.

All of these competitors have been accepted by ITU as part of the IMT-2000 family of 3G standards, along with UMTS-FDD.

On the Internet access side, competing systems include WiMAX and Flash-OFDM.

Migrating from GSM/GPRS to UMTS

From a GSM/GPRS network, the following network elements can be reused:

* Home Location Register (HLR).

- Visitor Location Register (VLR).

- Equipment Identity Register (EIR).

- Mobile Switching Center (MSC) (vendor dependent).

- Authentication Center (AUC).

- Serving GPRS Support Node (SGSN) (vendor dependent).

- Gateway GPRS Support Node (GGSN).

From a GSM/GPRS communication radio network, the following elements cannot be reused:

- Base station controller (BSC).

- Base transceiver station (BTS).

They can remain in the network and be used in dual network operation where 2G and 3G networks co-exist while network migration and new 3G terminals become available for use in the network.

The UMTS network introduces new network elements that function as specified by 3GPP:

- Node B (base transceiver station).

- Radio Network Controller (RNC).

- Media Gateway (MGW).

The functionality of MSC and SGSN changes when going to UMTS. In a GSM system the MSC handles all the circuit switched operations like connecting A- and B-subscriber through the network. SGSN handles all the packet switched operations and transfers all the data in the network. In UMTS the Media gateway (MGW) take care of all data transfer in both circuit and packet switched networks. MSC and SGSN control MGW operations. The nodes are renamed to MSC-server and GSN-server.

Problems and Issues

Some countries, including the United States, have allocated spectrum differently from the ITU recommendations, so that the standard bands most commonly used for UMTS (UMTS-2100) have not been available. In those countries, alternative bands are used, preventing the interoperability of existing UMTS-2100 equipment, and requiring the design and manufacture of different equipment for the use in these markets. As is the case with GSM900 today, standard UMTS 2100 MHz equipment will not work in those markets. However, it appears as though UMTS is not suffering as much from handset band compatibility issues as GSM did, as many UMTS handsets are multi-band in both UMTS and GSM modes. Penta-band (850, 900, 1700, 2100, and 1900 MHz bands), quad-band GSM (850, 900, 1800, and 1900 MHz bands) and tri-band UMTS (850, 1900, and 2100 MHz bands) handsets are becoming more commonplace.

In its early days, UMTS had problems in many countries: Overweight handsets with poor battery life were first to arrive on a market highly sensitive to weight and form factor. The Motorola A830, a debut handset on Hutchison's 3 network, weighed more than 200 grams and even featured a detachable camera to reduce handset weight. Another significant issue involved call reliability, related to problems with handover from UMTS to GSM. Customers found their connections being dropped as handovers were possible only in one direction (UMTS → GSM), with the handset only changing back to UMTS after hanging up. In most networks around the world this is no longer an issue.

Compared to GSM, UMTS networks initially required a higher base station density. For fully-fledged UMTS incorporating video on demand features, one base station needed to be set up every 1–1.5 km (0.62–0.93 mi). This was the case when only the 2100 MHz band was being used, however with the growing use of lower-frequency bands (such as 850 and 900 MHz) this is no longer so. This has led to increasing rollout of the lower-band networks by operators since 2006.

Even with current technologies and low-band UMTS, telephony and data over UMTS requires more power than on comparable GSM networks. Apple Inc. cited UMTS power consumption as the reason that the first generation iPhone only supported EDGE. Their release of the iPhone 3G quotes talk time on UMTS as half that available when the handset is set to use GSM. Other manufacturers indicate different battery lifetime for UMTS mode compared to GSM mode as well. As battery and network technology improve, this issue is diminishing.

Security Issues

As early as 2008, it was known that carrier networks can be used to surreptitiously gather user location information. In August 2014, the Washington Post reported on widespread marketing of surveillance systems using Signalling System No. 7 (SS7) protocols to locate callers anywhere in the world.

In December 2014, news broke that SS7's very own functions can be repurposed for surveillance, because of its lax security, in order to listen to calls in real time or to record encrypted calls and texts for later decryption, or to defraud users and cellular carriers.

Deutsche Telekom and Vodafone declared the same day that they had fixed gaps in their networks, but that the problem is global and can only be fixed with a telecommunication system-wide solution.

Fourth Generation

4G is the fourth generation of broadband cellular network technology, succeeding 3G. A 4G system must provide capabilities defined by ITU in IMT Advanced. Potential and current applications include amended mobile web access, IP telephony, gaming services, high-definition mobile TV, video conferencing, and 3D television.

The first-release Long Term Evolution (LTE) standard was commercially deployed in Oslo, Norway, and Stockholm, Sweden in 2009, and has since been deployed throughout most parts of the world. It has, however, been debated whether first-release versions should be considered 4G LTE.

In March 2009, the International Telecommunications Union-Radio communications sector (ITU-R) specified a set of requirements for 4G standards, named the International Mobile Telecommunications Advanced (IMT-Advanced) specification, setting peak speed requirements for 4G service at 100 megabits per second (Mbit/s)(=12.5 megabytes per second) for high mobility communication (such as from trains and cars) and 1 gigabit per second (Gbit/s) for low mobility communication (such as pedestrians and stationary users).

Since the first-release versions of Mobile WiMAX and LTE support much less than 1 Gbit/s peak bit rate, they are not fully IMT-Advanced compliant, but are often branded 4G by service providers. According to operators, a generation of the network refers to the deployment of a new non-backward-compatible technology. On December 6, 2010, ITU-R recognized that these two technologies, as well as other beyond-3G technologies that do not fulfill the IMT-Advanced requirements, could nevertheless be considered "4G", provided they represent forerunners to IMT-Advanced compliant versions and "a substantial level of improvement in performance and capabilities with respect to the initial third generation systems now deployed".

Mobile WiMAX Release 2 (also known as *WirelessMAN-Advanced* or *IEEE 802.16m'*) and LTE Advanced (LTE-A) are IMT-Advanced compliant backwards compatible versions of the above two systems, standardized during the spring 2011, and promising speeds in the order of 1 Gbit/s. Services were expected in 2013.

As opposed to earlier generations, a 4G system does not support traditional circuit-switched telephony service, but instead relies on all-Internet Protocol (IP) based communication such as IP telephony. The spread spectrum radio technology used in 3G systems is abandoned in all 4G candidate systems and replaced by OFDMA multi-carrier transmission and other frequency-domain equalization (FDE) schemes, making it possible to transfer very high bit rates despite extensive multi-path radio propagation (echoes). The peak bit rate is further improved by smart antenna arrays for multiple-input multiple-output (MIMO) communications.

In the field of mobile communications, a "generation" generally refers to a change in the fundamental nature of the service, non-backwards-compatible transmission technology, higher peak bit rates, new frequency bands, wider channel frequency bandwidth in Hertz, and higher capacity for many simultaneous data transfers (higher system spectral efficiency in bit/second/Hertz/site).

New mobile generations have appeared about every ten years since the first move from 1981 analog (1G) to digital (2G) transmission in 1992. This was followed, in 2001, by 3G multi-media support, spread spectrum transmission and, at least, 200 kbit/s peak bit rate, in 2011/2012 to be followed by "real" 4G, which refers to all-Internet Protocol (IP) packet-switched networks giving mobile ultra-broadband (gigabit speed) access.

While the ITU has adopted recommendations for technologies that would be used for future global communications, they do not actually perform the standardization or development work them, instead relying on the work of other standard bodies such as IEEE, The Wi MAX Forum, and 3GPP.

In the mid-1990s, the ITU-R standardization organization released the IMT-2000 requirements as a framework for what standards should be considered 3G systems, requiring 200 kbit/s peak bit rate. In 2008, ITU -R specified the IMT – Advanced (International Telecommunications Advanced) requirements for 4G systems.

The fastest 3G-based standard in the UMTS family is the HSPA+ standard, which is commercially available since 2009 and offers 28 Mbit/s downstream (22 Mbit/s upstream) without MIMO, i.e. only with one antenna, and in 2011 accelerated up to 42 Mbit/s peak bit rate downstream using either DC-HSPA+ (simultaneous use of two 5 MHz UMTS carriers) or 2x2 MIMO. In theory speeds up to 672 Mbit/s are possible, but have not been deployed yet. The fastest 3G-based standard in the CDMA2000 family is the EV-DO Rev. B, which is available since 2010 and offers 15.67 Mbit/s downstream.

Frequencies for 4G LTE Networks

Mobile 4G network uses several frequencies:

- 700 MHz (Band 28 - Telstra/Optus).
- 850 MHz (Band 5 - Vodafone).
- 900 MHz (Band 8 - Telstra).
- 1800 MHz (Band 3 - Telstra/Optus/Vodafone).
- 2100 MHz (Band 1 - [a small number of Telstra sites]/Optus [Tasmania]/Vodafone).
- 2300 MHz (Band 40 - Optus [Vivid Wireless spectrum]).
- 2600 MHz (Band 7 - Telstra/Optus).

In Australia, the 700 MHz band was previously used for analogue television and became operational with 4G in December 2014. The 850 MHz band is currently operated as a 3G network by Telstra and as a 4G network by Vodafone in Australia.

IMT-Advanced Requirements

An IMT-Advanced cellular system must fulfill the following requirements:

- Be based on an all-IP packet switched network.
- Have peak data rates of up to approximately 100 Mbit/s for high mobility such as mobile access and up to approximately 1 Gbit/s for low mobility such as nomadic/local wireless access.
- Be able to dynamically share and use the network resources to support more simultaneous users per cell.
- Use scalable channel bandwidths of 5–20 MHz, optionally up to 40 MHz.
- Have peak link spectral efficiency of 15 bit/s·Hz in the downlink, and 6.75 bit/s·Hz in the up link (meaning that 1 Gbit/s in the downlink should be possible over less than 67 MHz bandwidth).
- System spectral efficiency is, in indoor cases, 3 bit/s·Hz·cell for downlink and 2.25 bit/s·Hz·-cell for uplink.
- Smooth handovers across heterogeneous networks.

In September 2009, the technology proposals were submitted to the International Telecommunication Union (ITU) as 4G candidates. Basically all proposals are based on two technologies:

- LTE Advanced standardized by the 3GPP.

- 802.16m standardized by the IEEE.

Implementations of Mobile WiMAX and first-release LTE are largely considered a stopgap solution that will offer a considerable boost until WiMAX 2 (based on the 802.16m specification) and LTE Advanced are deployed. The latter's standard versions were ratified in spring 2011, but are still far from being implemented.

The first set of 3GPP requirements on LTE Advanced was approved in June 2008. LTE Advanced was to be standardized in 2010 as part of Release 10 of the 3GPP specification. LTE Advanced will be based on the existing LTE specification Release 10 and will not be defined as a new specification series. A summary of the technologies that have been studied as the basis for LTE Advanced is included in a technical report.

Some sources consider first-release LTE and Mobile WiMAX implementations as pre-4G or near-4G, as they do not fully comply with the planned requirements of 1 Gbit/s for stationary reception and 100 Mbit/s for mobile.

Confusion has been caused by some mobile carriers who have launched products advertised as 4G but which according to some sources are pre-4G versions, commonly referred to as *3.9G*, which do not follow the ITU-R defined principles for 4G standards, but today can be called 4G according to ITU-R. Vodafone NL for example, advertised LTE as *4G*, while advertising now LTE Advanced as their '4G+' service which actually is (true) 4G. A common argument for branding 3.9G systems as new-generation is that they use different frequency bands from 3G technologies; that they are based on a new radio-interface paradigm; and that the standards are not backwards compatible with 3G, whilst some of the standards are forwards compatible with IMT-2000 compliant versions of the same standards.

System Standards

IMT-2000 compliant 4G standards

As of October 2010, ITU-R Working Party 5D approved two industry-developed technologies (LTE Advanced and WirelessMAN-Advanced) for inclusion in the ITU's International Mobile Telecommunications Advanced program (IMT-Advanced program), which is focused on global communication systems that will be available several years from now.

LTE Advanced

LTE Advanced (Long Term Evolution Advanced) is a candidate for IMT-Advanced standard, formally submitted by the 4GPP organization to ITU-T in the fall 2009, and expected to be released in 2013. The target of 3GPP LTE Advanced is to reach and surpass the ITU requirements. LTE Advanced is essentially an enhancement to LTE. It is not a new technology, but rather an improvement on the existing LTE network. This upgrade path makes it more cost effective for vendors to offer LTE and then upgrade to LTE Advanced which is similar to the upgrade from WCDMA to HSPA. LTE and LTE Advanced will also make use of additional spectrums and multiplexing to allow it to achieve higher data speeds. Coordinated Multi-point Transmission will also allow more system capacity to help handle the enhanced data speeds. Release 10 of LTE is expected to achieve the IMT Advanced speeds. Release 8 currently supports up to 300 Mbit/s of download speeds which is still short of the IMT-Advanced standards.

Table: Data speeds of LTE-Advanced.	
	LTE Advanced
Peak download	1000 Mbit/s
Peak upload	0500 Mbit/s

IEEE 802.16m or WirelessMAN-Advanced

The IEEE 802.16m or WirelessMAN-Advanced evolution of 802.16e is under development, with the objective to fulfill the IMT-Advanced criteria of 1 Gbit/s for stationary reception and 100 Mbit/s for mobile reception.

Forerunner Versions

3GPP Long Term Evolution (LTE)

Telia-branded Samsung LTE modem.

The pre-4G 3GPP Long Term Evolution (LTE) technology is often branded "4G – LTE", but the first LTE release does not fully comply with the IMT-Advanced requirements. LTE has a theoretical net bit rate capacity of up to 100 Mbit/s in the downlink and 50 Mbit/s in the uplink if a 20 MHz channel is used — and more if multiple-input multiple-output (MIMO), i.e. antenna arrays, are used.

The physical radio interface was at an early stage named *High Speed OFDM Packet Access* (HSO-PA), now named Evolved UMTS Terrestrial Radio Access (E-UTRA). The first LTE USB dongles do not support any other radio interface.

The world's first publicly available LTE service was opened in the two Scandinavian capitals, Stockholm (Ericsson and Nokia Siemens Networks systems) and Oslo (a Huawei system) on December

14, 2009, and branded 4G. The user terminals were manufactured by Samsung. As of November 2012, the five publicly available LTE services in the United States are provided by MetroPCS, Verizon Wireless, AT&T Mobility, U.S. Cellular, Sprint, and T-Mobile US.

T-Mobile Hungary launched a public beta test (called *friendly user test*) on 7 October 2011, and has offered commercial 4G LTE services since 1 January 2012.

In South Korea, SK Telecom and LG U+ have enabled access to LTE service since 1 July 2011 for data devices, slated to go nationwide by 2012. KT Telecom closed its 2G service by March 2012, and complete the nationwide LTE service in the same frequency around 1.8 GHz by June 2012.

In the United Kingdom, LTE services were launched by EE in October 2012, by O2 and Vodafone in August 2013, and by Three in December 2013.

Table: Data speeds of LTE.	
	LTE
Peak download	0100 Mbit/s
Peak upload	0050 Mbit/s

Mobile WiMAX (IEEE 802.16e)

The Mobile WiMAX (IEEE 802.16e-2005) mobile wireless broadband access (MWBA) standard (also known as WiBro in South Korea) is sometimes branded 4G, and offers peak data rates of 128 Mbit/s downlink and 56 Mbit/s uplink over 20 MHz wide channels.

In June 2006, the world's first commercial mobile WiMAX service was opened by KT in Seoul, South Korea.

Sprint has begun using Mobile WiMAX, as of 29 September 2008, branding it as a "4G" network even though the current version does not fulfill the IMT Advanced requirements on 4G systems.

In Russia, Belarus and Nicaragua WiMax broadband internet access were offered by a Russian company Scartel, and was also branded 4G, Yota.

Table: Data speeds of WiMAX.	
	WiMAX
Peak download	0128 Mbit/s
Peak upload	0056 Mbit/s

In the latest version of the standard, WiMax 2.1, the standard has been updated to be not compatible with earlier WiMax standard, and is instead interchangeable with LTE-TDD system, effectively merging WiMax standard with LTE.

TD-LTE for China Market

Just as Long-Term Evolution (LTE) and WiMAX are being vigorously promoted in the global telecommunications industry, the former (LTE) is also the most powerful 4G mobile communications leading technology and has quickly occupied the Chinese market. TD-LTE, one of the two variants of the LTE air interface technologies, is not yet mature, but many domestic and international wireless carriers are, one after the other turning to TD-LTE.

IBM's data shows that 67% of the operators are considering LTE because this is the main source of their future market. The above news also confirms IBM's statement that while only 8% of the operators are considering the use of WiMAX, WiMAX can provide the fastest network transmission to its customers on the market and could challenge LTE.

TD-LTE is not the first 4G wireless mobile broadband network data standard, but it is China's 4G standard that was amended and published by China's largest telecom operator – China Mobile. After a series of field trials, is expected to be released into the commercial phase in the next two years. Ulf Ewaldsson, Ericsson's vice president said: "the Chinese Ministry of Industry and China Mobile in the fourth quarter of this year will hold a large-scale field test, by then, Ericsson will help the hand." But viewing from the current development trend, whether this standard advocated by China Mobile will be widely recognized by the international market is still debatable.

Discontinued Candidate Systems

UMB (formerly EV-DO Rev. C)

UMB (Ultra Mobile Broadband) was the brand name for a discontinued 4G project within the 3GPP2 standardization group to improve the CDMA2000 mobile phone standard for next generation applications and requirements. In November 2008, Qualcomm, UMB's lead sponsor, announced it was ending development of the technology, favouring LTE instead. The objective was to achieve data speeds over 275 Mbit/s downstream and over 75 Mbit/s upstream.

Flash-OFDM

At an early stage the Flash-OFDM system was expected to be further developed into a 4G standard.

iBurst and MBWA (IEEE 802.20) systems

The iBurst system (or HC-SDMA, High Capacity Spatial Division Multiple Access) was at an early stage considered to be a 4G predecessor. It was later further developed into the Mobile Broadband Wireless Access (MBWA) system, also known as IEEE 802.20.

Principal Technologies in all Candidate Systems

Key Features

The following key features can be observed in all suggested 4G technologies:

- Physical layer transmission techniques are as follows:

    ○ MIMO: To attain ultrahigh spectral efficiency by means of spatial processing including multi-antenna and multi-user MIMO.

- ○ Frequency-domain-equalization, for example multi-carrier modulation (OFDM) in the downlink or single-carrier frequency-domain-equalization (SC-FDE) in the uplink: To exploit the frequency selective channel property without complex equalization.

- ○ Frequency-domain statistical multiplexing, for example (OFDMA) or (single-carrier FDMA) (SC-FDMA, a.k.a. linearly precoded OFDMA, LP-OFDMA) in the uplink: Variable bit rate by assigning different sub-channels to different users based on the channel conditions.

- ○ Turbo principle error-correcting codes: To minimize the required SNR at the reception side.

- Channel-dependent scheduling: To use the time-varying channel.

- Link adaptation: Adaptive modulation and error-correcting codes.

- Mobile IP utilized for mobility.

- IP-based femtocells (home nodes connected to fixed Internet broadband infrastructure).

As opposed to earlier generations, 4G systems do not support circuit switched telephony. IEEE 802.20, UMB and OFDM standards lack soft-handover support, also known as cooperative relaying.

Multiplexing and Access Schemes

Recently, new access schemes like Orthogonal FDMA (OFDMA), Single Carrier FDMA (SC-FDMA), Interleaved FDMA, and Multi-carrier CDMA (MC-CDMA) are gaining more importance for the next generation systems. These are based on efficient FFT algorithms and frequency domain equalization, resulting in a lower number of multiplications per second. They also make it possible to control the bandwidth and form the spectrum in a flexible way. However, they require advanced dynamic channel allocation and adaptive traffic scheduling.

WiMax is using OFDMA in the downlink and in the uplink. For the LTE (telecommunication), OFDMA is used for the downlink; by contrast, Single-carrier FDMA is used for the uplink since OFDMA contributes more to the PAPR related issues and results in nonlinear operation of amplifiers. IFDMA provides less power fluctuation and thus requires energy-inefficient linear amplifiers. Similarly, MC-CDMA is in the proposal for the IEEE 802.20 standard. These access schemes offer the same efficiencies as older technologies like CDMA. Apart from this, scalability and higher data rates can be achieved.

The other important advantage of the above-mentioned access techniques is that they require less complexity for equalization at the receiver. This is an added advantage especially in the MIMO environments since the spatial multiplexing transmission of MIMO systems inherently require high complexity equalization at the receiver.

In addition to improvements in these multiplexing systems, improved modulation techniques are being used. Whereas earlier standards largely used Phase-shift keying, more efficient systems such as 64QAM are being proposed for use with the 3GPP Long Term Evolution standards.

IPv6 Support

Unlike 3G, which is based on two parallel infrastructures consisting of circuit switched and packet switched network nodes, 4G is based on packet switching *only*. This requires low-latency data transmission.

As IPv4 addresses are (nearly) exhausted, IPv6 is essential to support the large number of wireless-enabled devices that communicate using IP. By increasing the number of IP addresses available, IPv6 removes the need for network address translation (NAT), a method of sharing a limited number of addresses among a larger group of devices, which has a number of problems and limitations. When using IPv6, some kind of NAT is still required for communication with legacy IPv4 devices that are not also IPv6-connected.

As of June 2009, Verizon has posted Specifications that require any 4G devices on its network to support IPv6.

Advanced Antenna Systems

The performance of radio communications depends on an antenna system, termed smart or intelligent antenna. Recently, multiple antenna technologies are emerging to achieve the goal of 4G systems such as high rate, high reliability, and long range communications. In the early 1990s, to cater for the growing data rate needs of data communication, many transmission schemes were proposed. One technology, spatial multiplexing, gained importance for its bandwidth conservation and power efficiency. Spatial multiplexing involves deploying multiple antennas at the transmitter and at the receiver. Independent streams can then be transmitted simultaneously from all the antennas. This technology, called MIMO (as a branch of intelligent antenna), multiplies the base data rate by (the smaller of) the number of transmit antennas or the number of receive antennas. Apart from this, the reliability in transmitting high speed data in the fading channel can be improved by using more antennas at the transmitter or at the receiver. This is called *transmit* or *receive diversity*. Both transmit/receive diversity and transmit spatial multiplexing are categorized into the space-time coding techniques, which does not necessarily require the channel knowledge at the transmitter. The other category is closed-loop multiple antenna technologies, which require channel knowledge at the transmitter.

Open-wireless Architecture and Software-defined Radio (SDR)

One of the key technologies for 4G and beyond is called Open Wireless Architecture (OWA), supporting multiple wireless air interfaces in an open architecture platform.

SDR is one form of open wireless architecture (OWA). Since 4G is a collection of wireless standards, the final form of a 4G device will constitute various standards. This can be efficiently realized using SDR technology, which is categorized to the area of the radio convergence.

Disadvantages

4G introduces a potential inconvenience for those who travel internationally or wish to switch carriers. In order to make and receive 4G voice calls, the subscriber handset must not only have a matching frequency band (and in some cases require unlocking), it must also have the matching enablement settings for the local carrier and/or country. While a phone purchased from a given carrier can be expected to work with that carrier, making 4G voice calls on another carrier's network (including international roaming) may be impossible without a software update specific to the local carrier and the phone model in question, which may or may not be available (although fallback to 3G for voice calling may still be possible if a 3G network is available with a matching frequency band).

Beyond 4G Research

A major issue in 4G systems is to make the high bit rates available in a larger portion of the cell, especially to users in an exposed position in between several base stations. In current research, this issue is addressed by macro-diversity techniques, also known as group cooperative relay, and also by Beam-Division Multiple Access (BDMA).

Pervasive networks are an amorphous and at present entirely hypothetical concept where the user can be simultaneously connected to several wireless access technologies and can seamlessly move between them (See vertical handoff, IEEE 802.21). These access technologies can be Wi-Fi, UMTS, EDGE, or any other future access technology. Included in this concept is also smart-radio (also known as cognitive radio) technology to efficiently manage spectrum use and transmission power as well as the use of mesh routing protocols to create a pervasive network.

Fifth Generation

Fifth-generation wireless (5G) is the latest iteration of cellular technology, engineered to greatly increase the speed and responsiveness of wireless networks. With 5G, data transmitted over wireless broadband connections could travel at rates as high as 20 Gbps by some estimates - exceeding wireline network speeds - as well as offer latency of 1 ms or lower for uses that require real-time feedback. 5G will also enable a sharp increase in the amount of data transmitted over wireless systems due to more available bandwidth and advanced antenna technology.

In addition to improvements in speed, capacity and latency, 5G offers network management features, among them network slicing, which allows mobile operators to create multiple virtual networks within a single physical 5G network. This capability will enable wireless network connections to support specific uses or business cases and could be sold on an as-a-service basis. A self-driving car, for example, would require a network slice that offers extremely fast, low-latency connections so a vehicle could navigate in real time. A home appliance, however, could be connected via a lower-power, slower connection because high performance is not crucial. The internet of things (IoT) could use secure, data-only connections.

5G networks and services will be deployed in stages over the next several years to accommodate the increasing reliance on mobile and internet-enabled devices. Overall, 5G is expected to generate a variety of new applications, uses and business cases as the technology is rolled out.

How 5G Works

Wireless networks are composed of cell sites divided into sectors that send data through radio waves. Fourth-generation (4G) Long-Term Evolution (LTE) wireless technology provides the foundation for 5G. Unlike 4G, which requires large, high-power cell towers to radiate signals over longer distances, 5G wireless signals will be transmitted via large numbers of small cell stations located in places like light poles or building roofs. The use of multiple small cells is necessary because the millimetre wave spectrum - the band of spectrum between 30 GHz and 300 GHz that 5G relies on to generate high speeds - can only travel over short distances and is subject to interference from weather and physical obstacles, like buildings.

Previous generations of wireless technology have used lower-frequency bands of spectrum. To offset millimeter wave challenges relating to distance and interference, the wireless industry is also considering the use of lower-frequency spectrum for 5G networks so network operators could use spectrum they already own to build out their new networks. Lower-frequency spectrum reaches greater distances but has lower speed and capacity than millimeter wave, however.

What is the Status of 5G Deployment?

Wireless network operators in four countries - the United States, Japan, South Korea and China - are largely driving the first 5G buildouts. Network operators are expected to spend billions of dollars on 5G capital expenses through 2030, according to Technology Business Research Inc., although it is not clear how 5G services will generate a return on that investment. Evolving use cases and business models that take advantage of 5G's benefits could address operators' revenue concerns.

Simultaneously, standards bodies are working on universal 5G equipment standards. The 3rd Generation Partnership Project (3GPP) approved 5G New Radio (NR) standards in December 2017 and is expected to complete the 5G mobile core standard required for 5G cellular services. The 5G radio system is not compatible with 4G radios, but network operators that have purchased wireless radios recently may be able to upgrade to the new 5G system via software rather than buying new equipment.

With 5G wireless equipment standards almost complete and the first 5G-compliant smartphones and associated wireless devices commercially available in 2019, 5G use cases will begin to emerge between 2020 and 2025, according to Technology Business Research projections. By 2030, 5G services will become mainstream and are expected to range from the delivery of virtual reality (VR) content to autonomous vehicle navigation enabled by real-time communications (RTC) capabilities.

What types of 5G Wireless Services will be Available?

Network operators are developing two types of 5G services:

- 5G fixed wireless broadband services deliver internet access to homes and businesses without a wired connection to the premises. To do that, network operators deploy NRs in small cell sites near buildings to beam a signal to a receiver on a rooftop or a windowsill that is amplified within the premises. Fixed broadband services are expected to make it less expensive for operators to deliver broadband services to homes and businesses because this approach eliminates the need to roll out fiber-optic lines to every residence. Instead, operators need only install fiber optics to cell sites, and customers receive broadband services through wireless modems located in their residences or businesses.

- 5G cellular services will provide user access to operators' 5G cellular networks. These services will begin to be rolled out in 2019 when the first 5G-enabled (or -compliant) devices are expected to become commercially available. Cellular service delivery is also dependent upon the completion of mobile core standards by 3GPP.

5G vs. 4G

Each generation of cellular technology is separated by not just their data transmission speed, but also a break in encoding methods which requires end-users to upgrade their hardware. 4G can support up to 2Gbps and are slowly continuing to improve in speeds. 4G featured speeds up to 500 times faster than 3G. 5G can be up to 100 times faster than 4G.

The main difference between 4 and 5G is the level of latency, of which 5G, will have much lower of. 5G will use OFDM encoding, similar to 4G LTE. 4G, however, will use 20 MHz channels; bonded together at 160 MHz. 5G will be up to between 100-800MHz channels, which require larger blocks of airwaves than 4G.

Samsung is currently researching into 6G. Not too much is currently known on how fast 6G would be and how it would operate; however, 6G will probably operate in similar magnitudes more than the differences between 4 and 5G. Some think 6G may use millimeter waves on the radio spectrum and may be a decade away.

References

- What-is-mobile-radio-telephone-system-or-og-and-what-is-0-5g, technology: cleardoubts.com, Retrieved 30 April, 2019
- "Htc - touch phone, pda phone, smartphone, mobile computer". Web.archive.org. 22 november 2008. Retrieved 17 august 2019
- What-is-1g-or-first-generation-of-wireless-telecommunication-technology, technology: cleardoubts.com, Retrieved 29 March, 2019
- 2g-technology-uses-features-advantages-and-disadvantages, technology: online-sciences.com, Retrieved 19 April, 2019
- 5g, definition: searchnetworking.techtarget.com, Retrieved 16 January, 2019

Mobile Network Architecture

- **Cell Tower**

- **Base Transceiver Station**

- **Cell Phone Signal Booster**

- **Base Station Subsystem**

- **Cellular Repeater**

- **GSM Network Architecture**

- **GPRS Architecture**

- **LTE Network Architecture**

Mobile network architecture consists of different elements which are necessary for mobile communication. Cell tower, base transceiver station, cell phone signal booster, base station subsystem, cellular repeater, etc. are some of the elements that fall under its domain. This chapter delves into different elements of mobile network architecture which will provide an easy understanding of the subject.

Cell Tower

A cell tower or also known as a cell site is a cellular telephone site where electric communications equipment and antennae are mounted. Some of this equipment are transmitters, receivers, control electronics, and additional power sources used for backup.

Cell towers are usually built by a tower company or a wireless carrier looking to expand their network coverage or capacity. The purpose of a cell phone tower is to facilitate the signal reception of cellular phones and other wireless communication devices like telephone, television, and radio in a cellular network.

How do Cell Towers Work?

Whenever you use your mobile phone to make a call, it emits electromagnetic radio waves also known as radio frequency or RF energy. Once the radio waves are emitted, the antenna from the nearest cell phone tower will receive them. The antennas of a cell tower can both transmit and receive signals from mobile phones. After receiving the radio waves from a mobile phone, it will then transmit the signals to a switching center which is a telephone exchange for mobile phones. This allows the call to be connected to either another mobile phone or to a telephone network.

Parts of a Cell Tower

A cell phone tower is made up of the following parts:

The Tower

There are four different types of cell towers:

- Lattice Tower: This is also known as a self-supporting tower. It affords the greatest flexibility and is often used in heavy loading conditions. It usually has three or four sides with similar shaped bases.

- Monopole Tower: This is a single steel or concrete tube tower. It usually does not exceed 50 meters and it requires one foundation. The antennas are attached on the exterior of the tower.

- Guyed Tower: This used to be the cheapest tower to construct. However, it requires the greatest amount of land. It's much cheaper to build a guyed tower for taller heights such as 100 meters or greater. Most of the radio and television towers are guyed towers. It is a straight tower connected by guy wires that anchor and support the tower. They are attached to the ground in all directions.

- Stealth Tower: These towers are typically required by councils and owners. They are more expensive than the other types of towers because they require additional material to stealth or hide their appearance. They also do not provide the same amount of capacity for tenants.

The Equipment

The tenants who attach their antennae on the cell tower uses transceivers and other supporting equipment installed in cabinets or in shelters. Wireless carriers have their own ways of protecting their equipment. Some places outdoor cabinets on concrete pads while others used prefabricated equipment shelters.

The Antennas

There are multiple antennas attached on a cell tower and they are typically mounted on a head frame. Sometimes only a few as three antennas are mounted while some cell towers have as many as fifteen antennas per carrier. The number of antennas depends on the technology, antenna performance, coverage, and capacity required.

Utilities

Cell towers have utilities installed at the site to be used by the carriers. Each initial carrier usually has power run to the site as well as phone service.

Access

Each cell tower or cell site will require access by the carriers for both initial installation and ongoing maintenance or repair activities. These may require a separate track to the mobile tower site.

The Cell Tower's Range

There are places where we often see a lot of cell towers. If they are allowed to stand near one another, what could the range of each cell tower be? The range of a cell tower in which mobile devices connects reliably to it is not a fixed figure and it may depend on some of these factors:

- The height of the antenna over the surrounding landscape.

- The frequency of the signal in use.

- The rated power of the transmitter.

- The directional characteristics of the antenna array on the site.

- The absorption and reflection of radio energy by buildings or vegetation.

- The local geographical or regulatory factors and weather conditions.

In those areas where there are enough cell towers to cover a wide area, the range of each one will be set to ensure that there is enough overlap from or to other sites and to ensure that the overlap area is not too large, to avoid interference problems with other cell towers or sites. Cell towers are usually grouped in areas with high population density or those places with the most potential users.

There are times when our mobile phones do not have a signal. It can be because we're too far from a tower, or the cell phone signals in our location are decreased by thick building walls, hills, or other structures. The signals from a cell tower do not need a clear line of sight but a greater radio interference will eliminate reception. Also, when a lot of people try to use a cell tower at the same time such as during events, or in a traffic jam, there's a tendency for the signal to be blocked.

When we're on the road or traveling somewhere, the base station controller and the intelligence of the cellphone keeps track of and allows our phones to switch from one cell tower to the next during the conversation. As the user moves towards a cell tower, it picks the strongest signal and releases the cell tower from which the signal has become weaker. The released channel from that cell tower will then become available to another user.

Cell towers are truly helpful in transmitting a signal to our mobile phones and other wireless devices. If you're having a difficult time finding a signal for your mobile phone, you might need to check if there are any structures blocking a cell tower from your location.

Femtocells

In telecommunications, a femtocell is a small, low-power cellular base station, typically designed for use in a home or small business. A broader term which is more widespread in the industry is small cell, with femtocell as a subset. It is also called femto AccessPoint (AP). It connects to the service provider's network via broadband (such as DSL or cable); current designs typically support four to eight simultaneously active mobile phones in a residential setting depending on version number and femtocell hardware, and eight to sixteen mobile phones in enterprise settings. A femtocell allows service providers to extend service coverage indoors or at the cell edge, especially where access would otherwise be limited or unavailable. Although much attention is focused on WCDMA, the concept is applicable to all standards, including GSM, CDMA2000, TD-SCDMA, WiMAX and LTE solutions.

A Verizon and AT&T femtocell.

The use of femtocells allows network coverage in places where the signal to the main network cells might be too weak. Furthermore, femtocells lower contention on the main network cells, by forming a connection from the end user, through an internet connection, to the operator's private network infrastructure elsewhere. The lowering of contention to the main cells plays a part in breathing, where connections are offloaded based on physical distance to cell towers.

Consumers and small businesses benefit from greatly improved coverage and signal strength since they have a *de facto* base station inside their premises. As a result of being relatively close to the femtocell, the mobile phone (user equipment) expends significantly less power for communication with it, thus increasing battery life. They may also get better voice quality (via HD voice) depending on a number of factors such as operator/network support, customer contract/price plan, phone and operating system support. Some carriers may also offer more attractive tariffs, for example discounted calls from home.

Femtocells are an alternative way to deliver the benefits of fixed–mobile convergence (FMC). The distinction is that most FMC architectures require a new dual-mode handset which works with existing unlicensed spectrum home/enterprise wireless access points, while a femtocell-based deployment will work with existing handsets but requires the installation of a new access point that uses licensed spectrum.

Many operators worldwide offer a femtocell service, mainly targeted at businesses but also offered to individual customers (often for a one-off fee) when they complain to the operator regarding a poor or non-existent signal at their location. Operators who have launched a femtocell service include SFR, AT&T, C Spire, Sprint Nextel, Verizon, Zain, Mobile TeleSystems, T-Mobile US, Orange, Vodafone, EE, O2, Three, and others.

In 3GPP terminology, a Home NodeB (HNB) is a 3G femtocell. A Home eNodeB (HeNB) is an LTE 4G femtocell.

Theoretically the range of a standard base station may be up to 35 kilometres (22 mi), and in practice could be 5–10 km (3–6 mi), a microcell is less than two kilometers wide, a picocell is 200 meters or less, and a femtocell is in the order of 10 meters, although AT&T calls its product, with a range of 40 feet (12 m), a "microcell". AT&T uses "AT&T 3G MicroCell" as a trademark and not necessarily the "microcell" technology, however.

Operating Mode

Femtocells are sold or loaned by a mobile network operator (MNO) to its residential or enterprise customers. A femtocell is typically the size of a residential gateway or smaller, and connects to the user's broadband line. Integrated femtocells (which include both a DSL router and femtocell) also exist. Once plugged in, the femtocell connects to the MNO's mobile network, and provides extra coverage. From a user's perspective, it is plug and play, there is no specific installation or technical knowledge required—anyone can install a femtocell at home.

In most cases, the user must then declare which mobile phone numbers are allowed to connect to their femtocell, usually via a web interface provided by the MNO. This needs to be done only once. When these mobile phones arrive under coverage of the femtocell, they switch over from the macrocell (outdoor) to the femtocell automatically. Most MNOs provide a way for the user to know this has happened, for example by having a different network name appear on the mobile phone. All communications will then automatically go through the femtocell. When the user leaves the femtocell coverage (whether in a call or not) area, their phone hands over seamlessly to the macro network. Femtocells require specific hardware, so existing WiFi or DSL routers cannot be upgraded to a femtocell.

Once installed in a specific location, most femtocells have protection mechanisms so that a location change will be reported to the MNO. Whether the MNO allows femtocells to operate in a different location depends on the MNO's policy. International location change of a femtocell is not permitted because the femtocell transmits licensed frequencies which belong to different network operators in different countries.

Benefits for Users

The main benefits for an end user are the following:

- "5 bar" coverage when there is no existing signal or poor coverage.

- Higher mobile data capacity, which is important if the end-user makes use of mobile data on his or her mobile phone (may not be relevant to a large number of subscribers who instead use WiFi where femtocell is located).

- Depending on the pricing policy of the MNO, special tariffs at home can be applied for calls placed under femtocell coverage.

- For enterprise users, having femtos instead of DECT ("cordless" home) phones enables them to have a single phone, so a single contact list, etc.

- Improved battery life for mobile devices due to reduced transmitter–receiver distance.

- The battery draining issue of mobile operators can be eliminated by means of energy efficiency of the networks resulting in prolongation of the battery life of handsets.

- New applications and services can be created to enhance user experience or provide additional features:

  ○ In Connected car case, the use of Femtocells has been proposed as a safety feature.

Femtocells can be used to give coverage in rural areas.

Standardised Architectures

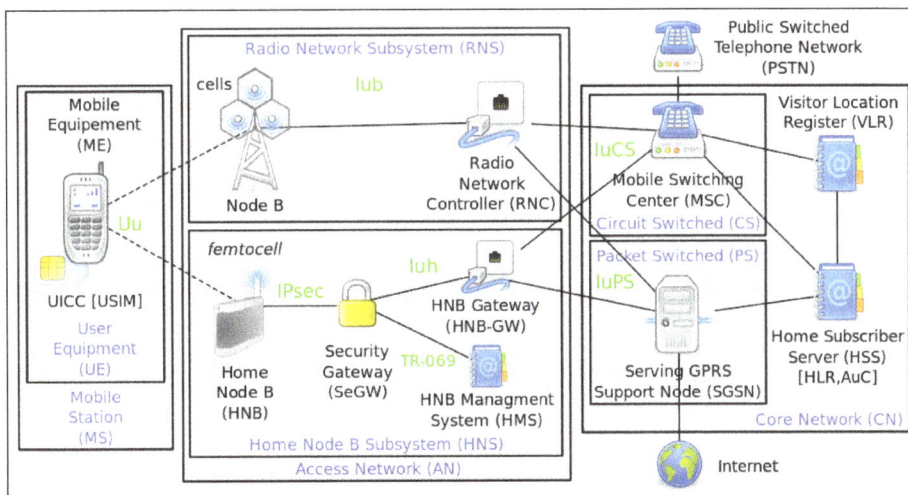

Simplified version of traditional Node B and Home Node B (3G femtocell) in 3G architecture.

The standards bodies have published formal specifications for femtocells for the most popular technologies, namely WCDMA, CDMA2000, LTE and WiMAX. These all broadly conform to an architecture with three major elements:

- The femtocell access points themselves, which embody greater network functionality than found in macrocell basestations, such as the radio resource control functions. This allows much greater autonomy within the femtocell, enabling self-configuration and self-optimisation. Femtocells are connected using broadband IP, such as DSL or cable modems, to the network operator's core switching centres.

- The femtocell gateway, comprising a security gateway that terminates large numbers of encrypted IP data connections from hundreds of thousands of femtocells, and a signalling gateway which aggregates and validates the signalling traffic, authenticates each femtocell and interfaces with the mobile network core switches using standard protocols, such as Luh.

- The management and operational system which allows software updates and diagnostic checks to be administered. These typically use the same TR-069 management protocol published by the Broadband Forum and also used for administration of residential modems.

The key interface in these architectures is that between the femtocell access points and the femtocell gateway. Standardisation enables a wider choice of femtocell products to be used with any gateway, increasing competitive pressure and driving costs down. For the common WCDMA femtocells, this is defined as the Luh interface. In the Luh architecture, the femtocell gateway sits between the femtocell and the core network and performs the necessary translations to ensure the femtocells appear as a radio network controller to existing mobile switching centres (MSCs). Each femtocell talks to the femtocell gateway and femtocell gateways talk to the Core Network Elements (CNE) (MSC for circuit-switched calls, SGSN for packet-switched calls). This model was proposed by 3GPP and the Femto Forum. New protocols (HNBAP [Home Node B Application Part] and RUA [RANAP User Adaptation]) have been derived; HNBAP is used for the control signaling between the HNB and HNB-GW while RUA is a lightweight mechanism to replace the SCCP and M3UA protocols in the RNC; its primary function is transparent transfer of RANAP messages.

In March 2010, the Femto Forum and ETSI conducted the first Plugfest to promote interoperability of the Luh standard.

The CDMA2000 standard released in March 2010 differs slightly by adopting the Session Initiation Protocol (SIP) to set up a connection between the femtocell and a femtocell convergence server (FCS). Voice calls are routed through the FCS which emulates an MSC. SIP is not required or used by the mobile device itself. In the SIP architecture, the femtocell connects to a core network of the mobile operator that is based on the SIP/IMS architecture. This is achieved by having the femtocells behave toward the SIP/IMS network like a SIP/IMS client by converting the circuit-switched 3G signaling to SIP/IMS signaling, and by transporting the voice traffic over RTP as defined in the IETF standards.

Air Interfaces

Although much of the commercial focus seems to have been on the Universal Mobile Telecommunications System (UMTS), the concept is equally applicable to all air-interfaces. Indeed, the first commercial deployment was the CDMA2000 Airave in 2007 by Sprint.

Femtocells are also under development or commercially available for GSM, TD-SCDMA, WiMAX and LTE.

The H(e)NB functionality and interfaces are basically the same as for regular High Speed Packet Access (HSPA) or LTE base stations except few additional functions. The differences are mostly to support differences in access control to support closed access for residential deployment or open access for enterprise deployment, as well as handover functionality for active subscribers and cell selection procedures for idle subscribers. For LTE additional functionality was added in 3GPP Release 9 which is summarized in.

Issues

Interference

The placement of a femtocell has a critical effect on the performance of the wider network, and this is the key issue to be addressed for successful deployment. Because femtocells can use the same frequency bands as the conventional cellular network, there has been the worry that rather than improving the situation they could potentially cause problems.

Femtocells incorporate interference mitigation techniques—detecting macrocells, adjusting power and scrambling codes accordingly. Ralph de la Vega, AT&T President, reported in June 2011 they recommended against using femtocells where signal strength was middle or strong because of interference problems they discovered after widescale deployment. This differs from previous opinions expressed by AT&T and others.

A good example is the comments made by Gordon Mansfield, Executive Director of RAN Delivery, AT&T, speaking at the Femtozone at CTIA March 2010:

"We have deployed femtocells co-carrier with both the hopping channels for GSM macrocells and with UMTS macrocells. Interference isn't a problem. We have tested femtocells extensively in real customer deployments of many thousands of femtocells, and we find that the mitigation techniques implemented successfully minimise and avoid interference. The more femtocells you deploy, the more uplink interference is reduced."

The Femto Forum has some extensive reports on this subject, which have been produced together with 3GPP and 3GPP2.

The simulations performed in the Femto Forum WG2 and 3GPP RAN4 encompass a wide spectrum of possible deployment scenarios including shared channel and dedicated channel deployments. In addition, the studies looked at the impact in different morphologies, as well as in closed versus open access. The following are broad conclusions from the studies:

- When femtocells are used in areas of poor or no coverage, macro/femto interference is unlikely to be a problem.

- If the femto network is sharing the channel (co-channel) with the macro network, interference can occur. However, if the interference management techniques advocated by the Femto Forum are adopted, the resulting interference can be mitigated in most cases.

- A femtocell network deployed on an adjacent dedicated channel is unlikely to create interference to a macro network. Additionally, the impact of a macro network on the performance of a femtocell on an adjacent channel is limited to isolated cases. If the interference mitigation techniques advocated by the Femto Forum are used, the impact is further marginalised.

- Closed access represents the worst-case scenario for creation of interference. Open access reduces the chances of User Equipment (mobile phone handsets, 3G data dongles, etc.) on the macro network interfering with a proximate femtocell.

- The same conclusions were reached for both the 850 MHz (3GPP Band 17) and 2100 MHz (3GPP Band 1) deployments that were studied.

The conclusions are common to the 850 MHz and 2100 MHz bands that were simulated in the studies, and can be extrapolated to other mobile bands. With interference mitigation techniques successfully implemented, simulations show that femtocell deployments can enable very high capacity networks by providing between a 10 and 100 times increase in capacity with minimal dead-zone impact and acceptable noise rise.

Femtocells can also create a much better user experience by enabling substantially higher data rates than can be obtained with a macro network and net throughputs that will be ultimately limited by backhaul in most cases (over 20 Mbps in 5 MHz).

Lawful Interception

Access point base stations, in common with all other public communications systems, are, in most countries, required to comply with lawful interception requirements.

Equipment Location

Other regulatory issues relate to the requirement in most countries for the operator of a network to be able to show exactly where each base-station is located, and for E911 requirements to provide the registered location of the equipment to the emergency services. There are issues in this regard for access point base stations sold to consumers for home installation, for example. Further, a consumer might try to carry his base station with him to a country where it is not licensed. Some manufacturers are using GPS within the equipment to lock the femtocell when it is moved to a different country; this approach is disputed, as GPS is often unable to obtain position indoors because of weak signal.

Emergency Calls

Access Point Base Stations are also required, since carrying voice calls, to provide a 911 (or 999, 112, etc.) emergency service, as is the case for VoIP phone providers in some jurisdictions. This service must meet the same requirements for availability as current wired telephone systems. Simply the phones must work if the AC mains grid is blacked out. There are several ways to achieve this, such as alternative power sources or fall-back to existing telephone infrastructure.

Quality of Service

When using an Ethernet or ADSL home backhaul connection, an Access Point Base Station must either share the backhaul bandwidth with other services, such as Internet browsing, gaming consoles, set-top boxes and triple-play equipment in general, or alternatively directly replace these functions within an integrated unit. In shared-bandwidth approaches, which are the majority of designs currently being developed, the effect on quality of service may be an issue.

The uptake of femtocell services will depend on the reliability and quality of both the cellular operator's network and the third-party broadband connection, and the broadband connection's subscriber understanding the concept of bandwidth utilization by different applications a subscriber may use. When things go wrong, subscribers will turn to cellular operators for support even if the root cause of the problem lies with the broadband connection to the home or workplace. Hence, the effects of any third-party ISP broadband network issues or traffic management policies need to be very closely monitored and the ramifications quickly communicated to subscribers.

A key issue recently identified is active traffic shaping by many ISPs on the underlying transport protocol IPSec.

Spectrum Accuracy

To meet Federal Communications Commission (FCC) / Ofcom spectrum mask requirements, femtocells must generate the radio frequency signal with a high degree of precision. To do this over a long period of time is a major technical challenge. The solution to this problem is to use an external, accurate signal to constantly calibrate the oscillator to ensure it maintains its accuracy. This is not simple (broadband backhaul introduces issues of network jitter/wander and recovered clock accuracy), but technologies such as the IEEE 1588 time synchronisation standard may address the issue. Also, Network Time Protocol (NTP) is being pursued by some developers as a possible solution to provide frequency stability. Conventional (macrocell) base stations often use GPS timing for synchronization and this could be used, although there are concerns on cost and the difficulty of ensuring good GPS coverage.

Standards bodies have recognized the challenge of this and the implications on device cost. For example, 3GPP has relaxed the 50ppb parts per billion precision to 100ppb for indoor base stations in Release 6 and a further loosening to 250ppb for Home Node B in Release 8.

Security

At the 2013 Black Hat hacker conference in Las Vegas, NV, a trio of security researchers detailed their ability to use a Verizon femtocell to secretly intercept the voice calls, data, and SMS text messages of any handset that connects to the device.

During a demonstration of their exploit, they showed how they could begin recording audio from a cell phone even before the call began. The recording included both sides of the conversation. They also demonstrated how it could trick Apple's iMessage – which encrypts texts sent over its network using SSL, rendering them unreadable to snoopers, to SMS, allowing the femtocell to intercept the messages.

They also demonstrated it was possible to "clone" a cell phone that runs on a CDMA network by remotely collecting its device ID number through the femtocell, in spite of added security measures to prevent against cloning of CDMA phones.

Controversy on Consumer Proposition

The impact of a femtocell is most often to improve cellular coverage, without the cellular carrier needing to improve their infrastructure (cell towers, etc.). This is net gain for the cellular carrier. However, the user must provide and pay for an internet connection to route the femtocell traffic, and then (usually) pay an additional one-off or monthly fee to the cellular carrier. Some have objected to the idea that consumers are being asked to pay to help relieve network shortcomings. On the other hand, residential femtocells normally provide a 'personal cell' which provides benefits only to the owner's family and friends.

The difference is also that while mobile coverage is provided through subscriptions from an operator with one business model, a fixed fibre or cable may work with a completely different business model. For example, mobile operators may imply restrictions on services which an operator on a fixed may not. Also, WiFi connects to a local network such as home servers and media players. This network should possibly not be within reach of the mobile operator.

Picocells

A picocell is a small cellular base station typically covering a small area, such as in-building (offices, shopping malls, train stations, stock exchanges, etc.), or more recently in-aircraft. In cellular networks, picocells are typically used to extend coverage to indoor areas where outdoor signals do not reach well, or to add network capacity in areas with very dense phone usage, such as train stations or stadiums. Picocells provide coverage and capacity in areas difficult or expensive to reach using the more traditional macrocell approach.

In cellular wireless networks, such as GSM, the picocell base station is typically a low-cost, small (typically the size of a ream of A4 paper), reasonably simple unit that connects to a base station controller (BSC). Multiple picocell 'heads' connect to each BSC: the BSC performs radio resource management and hand-over functions, and aggregates data to be passed to the mobile switching centre (MSC) or the gateway GPRS support node (GGSN).

Connectivity between the picocell heads and the BSC typically consists of in-building wiring. Although originally deployed systems (1990s) used plesiochronous digital hierarchy (PDH) links such as E1/T1 links, more recent systems use Ethernet cabling. Aircraft use satellite links.

More recent work has developed the concept towards a head unit containing not only a picocell, but also many of the functions of the BSC and some of the MSC. This form of picocell is sometimes called an access point base station or 'enterprise femtocell'. In this case, the unit contains all the capability required to connect directly to the Internet, without the need for the BSC/MSC infrastructure. This is a potentially more cost-effective approach.

Picocells offer many of the benefits of "small cells" (similar to femtocells) in that they improve data throughput for mobile users and increase capacity in the mobile network. In particular, the integration of picocells with macrocells through a heterogeneous network can be useful in seamless hand-offs and increased mobile data capacity.

Picocells are available for most cellular technologies including GSM, CDMA, UMTS and LTE from manufacturers including ip.access, ZTE, Huawei and Airwalk.

Range

Typically the range of a microcell is less than two kilometers wide, a picocell is 200 meters or less, and a femtocell is on the order of 10 meters, although AT&T calls its product, with a range of 40 feet (12 m), a "microcell". AT&T uses "AT&T 3G MicroCell" as a trademark and not necessarily the "microcell" technology, however.

Microcells

A microcell is a cell in a mobile phone network served by a low power cellular base station (tower), covering a limited area such as a mall, a hotel, or a transportation hub. A microcell is usually larger than a picocell, though the distinction is not always clear. A microcell uses power control to limit the radius of its coverage area.

Typically the range of a microcell is less than two kilometers wide, whereas standard base stations may have ranges of up to 35 kilometres (22 mi). A picocell, on the other hand, is 200 meters or less, and a femtocell is on the order of 10 meters, although AT&T calls its femtocell that has a range of 40 feet (12 m), a "microcell". AT&T uses "AT&T 3G MicroCell" as a trade mark and not necessarily the "microcell" technology, however.

A microcellular network is a radio network composed of microcells.

Rationale

Like picocells, microcells are usually used to add network capacity in areas with very dense phone usage, such as train stations. Microcells are often deployed temporarily during sporting events and other occasions in which extra capacity is known to be needed at a specific location in advance.

Cell size flexibility is a feature of 2G (and later) networks and is a significant part of how such networks have been able to improve capacity. Power controls implemented on digital networks make it easier to prevent interference from nearby cells using the same frequencies. By subdividing cells, and creating more cells to help serve high density areas, a cellular network operator can optimize the use of spectrum and ensure capacity can grow. By comparison, older analog systems have fixed limits beyond which attempts to subdivide cells simply would result in an unacceptable level of interference.

Microcell/Picocell-only Networks

Certain mobile phone systems, notably PHS and DECT, only provide microcellular (and Pico cellular) coverage. Microcellular systems are typically used to provide low cost mobile phone systems in high-density environments such as large cities. PHS is deployed throughout major cities in Japan as an alternative to ordinary cellular service. DECT is used by many businesses to deploy private license-free microcellular networks within large campuses where wireline phone service is less useful. DECT is also used as a private, non-networked, cordless phone system where its low power profile ensures that nearby DECT systems do not interfere with each other.

A forerunner of these types of network was the CT2 cordless phone system, which provided access to a looser network (without handover), again with base stations deployed in areas where large numbers of people might need to make calls. CT2's limitations ensured the concept never took off. CT2's successor, DECT, was provided with an interworking profile, GIP so that GSM networks could make use of it for microcellular access, but in practice the success of GSM within Europe, and the ability of GSM to support microcells without using alternative technologies, meant GIP was rarely used, and DECT's use in general was limited to non-GSM private networks, including use as cordless phone systems.

Macro Cells

A macrocell is a cell used in cellular networks with the function of providing radio coverage to a large area of mobile network access. A macrocell differs from a microcell by offering a larger coverage area and high-efficiency output. The macrocell is placed on stations where the output power is higher, usually in a range of tens of watts.

A macrocell is a radio coverage cell in cellular networks. The coverage distance varies depending on the frequency and bandwidth of the signals as well as physical obstructions in the area. Macrocell antennas, on the other hand must be properly mounted on ground-based masts, rooftops or other existing structures and at heights for an unhindered, clear view of the surroundings. Its performance can be increased by increasing the efficiency of the transceiver. Since this type of cell offers the largest coverage area, it is placed in stations along highways and rural areas where large stretches rarely have service within a few kilometers.

Base Transceiver Station

A typical BTS tower which holds the antenna. The tower is quite widely misinterpreted as the BTS itself. The shelter which houses the actual BTS can also be seen.

A mobile BTS.

A base transceiver station (BTS) is a piece of equipment that facilitates wireless communication between user equipment (UE) and a network. UEs are devices like mobile phones (handsets), WLL phones, computers with wireless Internet connectivity. The network can be that of any of the wireless communication technologies like GSM, CDMA, wireless local loop, Wi-Fi, WiMAX or other wide area network (WAN) technology.

BTS is also referred to as the *node B* (in 3G Networks) or, simply, the *Base Station* (BS).

Though the term BTS can be applicable to any of the wireless communication standards, it is generally associated with mobile communication technologies like GSM and CDMA. In this regard, a BTS forms part of the base station subsystem (BSS) developments for system management. It may also have equipment for encrypting and decrypting communications, spectrum filtering tools (band pass filters), etc. antennas may also be considered as components of BTS in general sense as they facilitate the functioning of BTS. Typically a BTS will have several transceivers (TRXs) which allow it to serve several different frequencies and different sectors of the cell (in the case of sectorised base stations). A BTS is controlled by a parent base station controller via the base station control function (BCF). The BCF is implemented as a discrete unit or even incorporated in a TRX in compact base stations. The BCF provides an operations and maintenance (O&M) connection to the network management system (NMS), and manages operational states of each TRX, as well as software handling and alarm collection. The basic structure and functions of the BTS remains the same regardless of the wireless technologies.

General Architecture

A BTS is usually composed of:

- Transceiver (TRX): Provides transmission and reception of signals. It also does sending and reception of signals to and from higher network entities (like the base station controller in mobile telephony).

- Power amplifier (PA): Amplifies the signal from TRX for transmission through antenna; may be integrated with TRX.

- Combiner: Combines feeds from several TRXs so that they could be sent out through a single antenna. Allows for a reduction in the number of antenna used.

- Multiplexer: For separating sending and receiving signals to/from antenna. Does sending and receiving signals through the same antenna ports (cables to antenna).

- Antenna: This is the structure that the BTS lies underneath; it can be installed as it is or disguised in some way (Concealed cell sites).

- Alarm extension system: Collects working status alarms of various units in the BTS and extends them to operations and maintenance (O&M) monitoring stations.

- Control function.

- Controls and manages the various units of BTS, including any software. On-the-spot configurations, status changes, software upgrades, etc. are done through the control function.

- Baseband receiver unit (BBxx).

- Frequency hopping, signal DSP.

Electronic equipment box at the base of cellphone tower at Chinawal village, India.

Diversity Techniques

To improve the quality of the received signal, often two receiving antennas are used, placed at a distance equal to an odd multiple of a quarter of the corresponding wavelength. For 900 MHz, this wavelength is 33 cm. This technique, known as antenna diversity or space diversity, avoids interruption caused by path fading. The antennas can be spaced horizontally or vertically. Horizontal spacing requires more complex installation, but brings better performance.

Other than antenna or space diversity, there are other diversity techniques such as frequency/time diversity, antenna pattern diversity, and polarization diversity.

Splitting refers to the flow of power within a particular area of the cell, known as a sector. Every field can therefore be considered like one new cell.

Directional antennas reduce LoRa (Long Range) interference. If not sectorised, the cell will be served by an omnidirectional antenna, which radiates in all directions. A typical structure is the trisector, also known as clover, in which there are three sectors served by separate antennas. Each sector has a separate direction of tracking, typically of 120° with respect to the adjacent ones. Other orientations may be used to suit the local conditions. Bisectored cells are also implemented. These are most often oriented with the antennas serving sectors of 180° separation to one another, but again, local variations do exist.

Cell Phone Signal Booster

A cell phone signal booster (also known as amplifier or repeater) is made up of three main elements - exterior antenna, amplifier, and interior antenna. They form a wireless system to boost cellular reception.

A cell phone reception booster is generally a repeater system that involves the amplifier adding gain or power to the reception in various directions. Even for a cheap cell phone signal booster, maximum gain differs by application. The work of an outside antenna is to both receive and transmit signal to a cellular tower with enhanced power and sensitivity. Usually the dB gain is never below 7 dB and can be over 10 dB gain. The system's elements conduit is the coaxial cable. It is also a factor in transmission loss.

The main aim of the cellular phone signal booster is taking the existing cell phone signal around your car, office, workstation or home and amplifying it. After amplification, the signal is rebroadcasted to the area with no reception or weak signal. Apart from consisting of an amplifier to boost the reception, external antenna and internal antenna, there're cell phone boosters with indoor antenna and amplifier as a single unit making them superb indoor cell phone signal boosters. In most cases the three components are detached. Other optional components include the Attenuator (reduces unwanted frequency signals), Lightning Surge Protector, Splitter and Tap.

How a Cell Phone Reception Booster Works

A Sprint cell phone signal booster and repeaters for other carriers come into use in places where the cell phone reception is weak. The strength of the cell phone signal is usually affected by diverse obstructions. This includes natural obstructions like trees, construction materials in buildings and hills as well as distance. This is where the reception booster system comes in. The mobile signal of a cellular phone refers to the strength of the signal that is calculated in dBm that a cellular network sends to the mobile phone.

Firstly, the cell phone reception is captured by the outside antenna. It is then amplified by the cellular repeater and rebroadcasted across the building or car through the inside antenna. The result is a boosted cell phone reception that culminates in more bars in your phone even in the remotest of places. You can then enjoy clearer calls (you won't drop calls again) and faster internet browsing and rapid data downloads and uploads. This is how the cellphone signal booster works in your car or house. This is how all signal boosters work on all technology platform networks including GSM, CDMA, TDMA, EVDO, UMTS, HSPA+, and the latest LTE.

All weBoost and Wilson repeaters, including cell phone signal booster T Mobile systems among others, work with virtually all cellular carriers in North America, particularly in Canada and United States. The boosters also work with all manner of cellular devices. Data and voice speeds from 2G voice, 3G data and 4G LTE advanced networks are all boosted depending on the booster you choose.

Another advantage of a boosted connection is that the cell phone battery life will be extended by 180 minutes of extra talk time. You can actually choose a mobile signal booster that amplifies the signal strength of diverse cell phones simultaneously.

Signal strength refers to the electric field magnitude at a point of reference, which is a bit distant from the antenna doing the transmission. Signal strength is also known as field strength or received level signal. As you talk on your phone, the voice is changed by the mobile device into electric signals and relayed as radio waves. The phone of the person you're talking to then converts the electrical signals into sound that the individual can hear. Any cell phone signal booster for house, car or marine vessel works to amplify this transmission.

Types of Cell Phone Repeater Reception Boosters

- Analog Signal Repeaters: Essentially, most amplifiers today are analog ones. They amplify all mobile phone carrier frequencies using traditional technology. Also broadband (wide band) amplifiers, they are usually sold together with an outdoor antenna cable and kit. Installation is needed. Sometimes analog signal amplifiers are known as Bi-Directional repeaters/amplifier or BDAs in short. This is a signal booster that lots of localities under their law require to allow first responders to remain in contact around different areas and facilities to cater for emergency situations. Bi-directional amplifiers are for both cellular phone and two-way radio use.

- Smart Signal Boosters: Generally, this defines a new form of wireless cell phone signal booster using very all-digital powerful baseband processors that clean the coverage prior to rebroadcasting. Smart reception amplifiers come with over 100 dB gains while their analog counterparts have 63-70 dB gains. They might be a little expensive than analog ones, but come with such great features as plug and play, internal donor antenna inside the box of the booster and no outdoor antennas are needed.

Reasons why your Signal is Weak

Testing the cellular coverage in an area full of 100 feet deep wood red and white pine in New York, a Wilson signal booster was able to overcome the various obstacles manifested in rural areas across America. The result was a reliable cell phone reception to those living in the home. Data and voice reception from various carriers was enhanced.

- Distance between the cellular tower and your vehicle/home: One of the reasons why your cell phone reception is weak could be due to the distance from the cellular tower closest to you. The closer you're to the cell tower, the stronger the cell phone signal received. On other hand, the farther you're from the cellular tower of your carrier, the worse the cell phone signal.

- Interferences from the outside: It is also possible external interferences are affecting your cellular coverage. Note that cellular signals are generally radio waves and obstructed a lot as they cover long distances to reach your phone. Effective transmission of the waves requires a clear line right to the carrier's tower. Nonetheless, such outside interruptions as mountains, trees, skyscrapers and other tall buildings, hills, billboards, elements such as snow, thunderstorms and rain weaken the reception.

- Interference from the inside: Another thing that weakens the reception is internal interferences. These include such things as thick building materials like layers of brick and thick concrete, radiant barriers, glasses and metal, magnetic and electronic clutter as well as conductive materials that block or weaken an incoming coverage. Your outside signal may be superb and perhaps even very close to your carrier cellular tower but inside your house the signal can be quite weak due to internal interferences.

Certification of Repeaters

Today, in both Canada and the United States, all the cell phone signal booster repeaters have to be certified. Certification by the two federal governments mean that the amplifiers work as required or indicated; deal with reduced cellular coverage or extend the range of the cell phone. FCC (Federal Communication Commission) is United States body mandated with the certification and testing of all the cellular boosters up for sale. Across the border, it is IC (Industry Canada) that has the responsibility of ensuring this happens in Canada.

What is an Aptitude of a Reception Booster?

An ATT cell phone signal booster and other carriers repeaters are made to work with an existing transmission. The transmission is then amplified accordingly before broadcasting it to a location or space with a very weak or no reception at all. Nonetheless, the system should work if there's already a stable signal that the device will be receiving and amplifying. Mobile repeaters, or any other repeater, don't make signals and only transmit and amplify the cellular carrier's signal. If the external cell phone antenna cannot be placed in an area or surrounding where signal is stable, the cell phone reception booster might not work as you desire. The right cell phone signal booster app can actually help identify such feasibility.

Strength of the Signal

The cell phone coverage of the location you intend to have the external antenna mounted will determine the total area that need to be covered, perhaps inside your vehicle, office or home. If external reception is strong enough, you can easily cover a section that is about the same size as the coverage area promoted on a repeater. On the other hand, if there's less external strong signal, interior area to be covered will not be the same size as that advertised. It might require going for a more powerful signal system to get a reliable coverage. Make a note of this any time you're buying a Wilson cell phone signal booster or for any other carrier, by paying attention to the square footage indicated.

You can also turn to the bars on a phone to evaluate the strength of the coverage. This will help you estimate the best place to have the external antenna mounted. There might be no universal way of showing signal bars on a mobile phone with some having five or less bars. Nonetheless, every bar is said to be 5-10 times stronger than the previous coverage bar. It means that those whose reception strength is very low (indicated on the cell phone as 1-2 bars) should not go for the recommended cell phone coverage booster for their location but one that is a little bit stronger.

Mind the Amplifier Power

Decibels (dBs) are a logarithmic type of measurement that rates the strength of an amplifier. If an amplifier is rated as promoting a 6 dB increase, it means its strength doubles. At the same time for every 6 dB decrease, the amplifier is half as powerful. If you're living around an area with a weak coverage and desire to enhance the connection across the entire area, choose a powerful amplifier. There's a circuit board in the booster box that amplifies signal received from input port, and sends it out from its output port.

A Cell Phone Reception Booster:

- Boosts the 4G LTE, 3G or 2G voice signals in your car or home.

- Enhances the internet speed, text and talk quality.

- Maintains a reliable connection throughout.

- Widens the reception area.

- Diminishes coverage problems.

On the other hand the cell phone reception booster cannot:

- Boost satellite and cable providers internet/Wi-Fi.

- Create working 4G LTE and 3G signals since it requires an existing coverage to work.

- Cannot work outside your vehicle or house but inside.

Outside Home/ Building Antennas

An outdoor cell phone signal booster like car cell phone booster usually features an outside antenna. An outdoor antenna is a critical component of any reception booster. It easily pulls in the weak cell phone coverage around you. Clearly, if your outside antenna is higher, the coverage pulled in is much better. They are easy to install or mount on a wall outside your house, on the house's roof or window.

Generally, the cell phone signal booster antenna includes a ground plane, a flat or almost flat horizontal that works as a surface for the conduction of radio waves from other elements of the antenna. Note that the ground planes don't have to be fixed on the ground. Their size and shape highly determine their gain and other radiation traits.

While purchasing Best Buy cell phone signal boosters for instance, you might want to think about a Yagi outside antenna or Omni directional outside antenna. With these antennas you can easily pull in the existing 4G LTE advanced or 3G signals.

An external antenna perhaps mounted on the house roof or via a window on the top story is connected with a coaxial cable to the base unit. The base unit then amplifies and relays the signal from the outdoors across the house.

- Omni-directional antenna: These are all-rounder antennas that will always perform. The cell phone reception is pulled from all the directions or 360 degrees. If you live in an area with lots of cell towers around and you intend to amplify multiple cellular carriers this is a wonderful antenna to think about. It supports various cellular service providers at the same time. You only need a moderate or weak signal and the Omni-directional antenna will basically do the rest.

- Yagi (Directional) Antenna: This uni-directional is a specialized antenna that pulls the cell phone reception from a specific direction, such as 45 degrees. It gives the antenna the chance to extend farther into the cellular tower for greater performance. Essentially, it is the kind of antenna to use for anyone living in the remotest of North American areas where there's a singular cellular tower or a single carrier. It is perfect for a poor signal with a big distance from the cellular tower and boosting the single network carrier is all that is needed.

Vehicle Outside Antennas

When it comes to cellular reception boosters for vehicles from RVs, trucks, marine vessels, motor homes, cars to SUVs there are diverse mounts and antennas to go for. Even so, you can choose between great car magnet standard mount type of antennas, superb large vehicle or recreational vehicle antennas and waterproof/weatherproof antennas for marine vessels for yachts or boats.

- Car Antenna: The typical car antenna is mounted vertically close to the middle of the roof of the car. As Omni-directional antennas, the cellular reception is pulled from diverse directions. It means the antenna should be clear of any obstruction.

- Most car antennas feature magnet mounts that are simple to install. The magnetic mount car antennas can be anything between 4 inches and 12 inches with a stable magnetic base that can stand on the metallic roof of a car. They are low profile antennas capable of pulling in 4G LTE and 3G coverage.

- Recreational Vehicle Antenna: A trucker or recreational vehicle antenna is a more durable and powerful specialized antenna. It is usually a bit long and rises between two and three feet. To absorb shock, the RV antenna is usually spring mounted with radial features for enhanced performance. It is not usually a magnetic antenna, and has to be fixed using an installation process with mounting assemblies that include screws, bolts, and metallic angled plates. While some RV antennas are specialized for 3G networks, most actually pull in both 4G LTE and 3G networks.

- Marine Antennas: Like the truck/RV antennas, these are also specialized. They are made with the weather and elements in mind due to the chaotic environment where they are used in high seas and rough oceans. Marine antennas would generally be under two feet in length and mostly made of salt-resistant material, fiberglass and stainless steel. They are also not magnetic but fixed installations and require a mount. Marine antennas usually pull in both 4G LTE and 3G networks.

Inside Antennae

A critical component of any cell phone reception booster is its inside antenna. This is vital because it receives the amplified transmission and extends or rebroadcasts it across the area the coverage is actually needed.

Building/Home inside Antennas

Dome and panel antennas serve inside buildings and home needs. Through splitters, multiple interior antennas can be paired with a single amplifier. They are great for long distance and multi-story coverage, or simply ensuring the reception is boosted in specific areas of a home or building.

- Panel inside antenna: This inside antenna effectively broadcasts the cell phone reception in a single direction. Panel antennas are wall mounted mostly and can also be mounted on ceilings to project reception below to lower levels/floors. If you have multiple floors to cover or wider rectangular spaces, these antennas will work fine.

- Dome inside antenna: This is a ceiling mounted antenna. It broadcasts cell phone reception in every direction and great for covering a single floor in a building.

Inside Antennas for Vehicles

To boost the cell phone reception in cars panel and low-profile inside antennas are the types used.

- Low profile car inside antenna: These antennas are optimal for sedans and cars. They come with a Velcro attachment for a stable mount. This antenna type is the one you will find in most vehicle cell phone reception boosters. Low profile inside car antennas relay boosted 4G LTE and 3G data and voice signals from the amplifier to the area within the vehicle. They broadcast the coverage up to a diameter of four feet.

- Panel inside antennas: Best for a large vehicle, panel inside antennas are best for use in boats and recreational vehicles. They boost voice signals, 4G LTE and 3G data signals in vehicles and cover more ground and room in a spacious marine vessel or large recreational vehicle or motor home.

Mind the Cables and Connectors

To connect antennas and amplifiers, coaxial cables are used. In a typical antenna booster reception booster kit, you will find two cable sets for connecting an amplifier to the antenna on the outside, and the other for connecting the inside antenna with amplifier placed indoors.

For flexibility in installation, a longer cable is best but it is wise to use least amount as possible because longer the wire/cable, more signal is lost during transmission within that cable. In essence, reception is diminished with a longer cable. This is why it is critical to work with the appropriate short cable grade to guarantee a reduced coverage loss.

- RG-6 Coaxial Cables: In homes, the F-connector RG-6 cables are heavily used. This is basically the typical connector and cable found in most homes with a cable or satellite television. They usually have 50 feet length coverage.

- LMR 400 Cable: While fixing a cell phone reception booster for a larger area the LMR 400 cable or its equivalent SureCall 400 is usually used due to large area coverage such as 10,000 square foot coverage. It is an ultralow loss type of cable with N-Connectors fitted on it. Compared to RG-6 coaxial cable, it is a bit thicker and comes in handy in areas where a longer cable will do. The cable length is usually up to 1000 feet.

- RG-174 Cable: For car cell phone reception boosters RG-174 coaxial cables are best. They come with SMA-connectors with a 10 foot cable length.

Booster Registration

Note that it is important today to register the cellular reception booster with the cellular carrier you're using. Of course, lots of wireless coverage providers have given consent for the use of virtually all approved reception cellular boosters certified by FCC that we carry. Nevertheless, as liability protection you should courteously register the reception booster to indicate compliance with fair use guidelines. Just contact the carrier and you will be furnished with the steps to complete registration.

What is the best cell phone signal booster to choose today? Are you searching for a portable cell phone signal booster for home and car? It all depends on your needs, for instance do you need the best cell phone antenna booster for use in your car, business premises and marine vessel or with your M2M applications? Of course, home cell phone signal booster reviews, for instance, can also help you to choose. Whether it is a cell phone signal amplifier Sprint carrier user or any other cellular carrier in Canada and USA there's a great zBoost repeater that meets your need. You can even check out our powerful commercial and industrial SureCall or Wilson cell phone signal boosters for better reception in your commercial building.

Cell Phone Signal Booster Home Uses

A commercial and home building cell phone reception amplifier, even a cell phone antenna amplifier from Best Buy will effectively enhance voice and 4G and 3G data networks. A phone booster is akin to having a personal cell tower in your own home or office building. Some of the best cell phone signal boosters for home include SureCall's SureCall EZ Call plug and play signal booster for voice, 3G, and 4G LTE which boosts reception over up to 2000 square feet. It is great for small properties, offices, small homes and small rooms with a 72 dB gain. Other in home cell phone signal boosters include SureCall Flare that boosts voice (2G), 3G and 4G LTE networks over up to 2500 square feet and boasts integrated booster and internal antenna for superior performance. Just for comparison with these home boosters, the Wilson Pro 4000R commercial amplifier for voice, 3G and 4G LTE networks is great for large businesses in gigantic spaces. It has a total coverage potential of 140,000 sq ft. Of course, cell phone signal booster for home reviews should shed more light on these repeater systems.

Cell Phone Signal Boosters Great for Vehicles

There're also cell phone signal boosters for cars, great for vehicles and marine vessels for 3G, 4G LTE and voice amplification. You can even access cell phone signal booster Radio Shack deals offering weBoost, SureCall, and Wilson repeaters. However, we recommend our robust SureCall Fusion2Go for cars/trucks and Fusion2Go-RV for recreational vehicles. Both are best used for their respective type of vehicle out there. Whether you want to build your own cell phone signal booster for your car/boat or not there're other better ones you can check out, such as multiuser boosters such as Fusion2Go-Boat for marine vessels such as boats, yachts and ships. They amplify voice, 3G and 4G Advanced LTE networks. Even T-mobile cell phone signal boosters, including other carriers for vehicles and marine vessels, are available as well. A phone signal booster review for vehicles will shed more light on this.

M2M Applications Signal Boosters

For machine to machine applications, there're signal boosters that work with all the various carrier services in Canada and USA. These boosters are best for machine to machine applications transmitting data wirelessly such as asset tracking systems, vending machines, remote monitoring systems, ATMs and fleet telematics among others. They include the part # SC-SoloAI-15 which is the 4G LTE M2M Signal Booster For Verizon And AT&T 4G LTE. These machine to machine signal boosters are direct connect amplifiers meaning they are wired to the cellular modem/device within an ATM, security panel cabinet, or vending machine where they require stronger cell connection.

Base Station Subsystem

The hardware of GSM base station displayed in Deutsches Museum.

The base station subsystem (BSS) is the section of a traditional cellular telephone network which is responsible for handling traffic and signaling between a mobile phone and the network switching subsystem. The BSS carries out transcoding of speech channels, allocation of radio channels to mobile phones, paging, transmission and reception over the air interface and many other tasks related to the radio network.

Base Transceiver Station

The base transceiver station, or BTS, contains the equipment for transmitting and receiving radio signals (transceivers), antennas, and equipment for encrypting and decrypting communications with the base station controller (BSC). Typically a BTS for anything other than a picocell will have several transceivers (TRXs) which allow it to serve several different frequencies and different sectors of the cell (in the case of sectorised base stations).

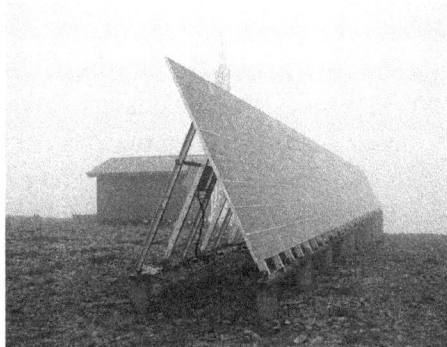

A solar-powered GSM base station on top of a
mountain in the wilderness of Lapland.

A BTS is controlled by a parent BSC via the "base station control function" (BCF). The BCF is implemented as a discrete unit or even incorporated in a TRX in compact base stations. The BCF provides an operations and maintenance (O&M) connection to the network management system (NMS), and manages operational states of each TRX, as well as software handling and alarm collection.

The functions of a BTS vary depending on the cellular technology used and the cellular telephone provider. There are vendors in which the BTS is a plain transceiver which receives information from the MS (mobile station) through the Um air interface and then converts it to a TDM (PCM) based interface, the Abis interface, and sends it towards the BSC. There are vendors which build their BTSs so the information is preprocessed, target cell lists are generated and even intracell handover (HO) can be fully handled. The advantage in this case is less load on the expensive Abis interface.

The BTSs are equipped with radios that are able to modulate layer 1 of interface Um; for GSM 2G+ the modulation type is Gaussian minimum-shift keying (GMSK), while for EDGE-enabled networks it is GMSK and 8-PSK. This modulation is a kind of continuous-phase frequency shift keying. In GMSK, the signal to be modulated onto the carrier is first smoothed with a Gaussian low-pass filter prior to being fed to a frequency modulator, which greatly reduces the interference to neighboring channels (adjacent-channel interference).

Antenna combiners are implemented to use the same antenna for several TRXs (carriers), the more TRXs are combined the greater the combiner loss will be. Up to 8:1 combiners are found in micro and pico cells only.

Frequency hopping is often used to increase overall BTS performance; this involves the rapid switching of voice traffic between TRXs in a sector. A hopping sequence is followed by the TRXs and handsets using the sector. Several hopping sequences are available, and the sequence in use for a particular cell is continually broadcast by that cell so that it is known to the handsets.

A TRX transmits and receives according to the GSM standards, which specify eight TDMA timeslots per radio frequency. A TRX may lose some of this capacity as some information is required to be broadcast to handsets in the area that the BTS serves. This information allows the handsets to identify the network and gain access to it. This signalling makes use of a channel known as the Broadcast Control Channel (BCCH).

Sectorization

By using directional antennas on a base station, each pointing in different directions, it is possible to sectorise the base station so that several different cells are served from the same location. Typically these directional antennas have a beamwidth of 65 to 85 degrees. This increases the traffic capacity of the base station (each frequency can carry eight voice channels) whilst not greatly increasing the interference caused to neighboring cells (in any given direction, only a small number of frequencies are being broadcast). Typically two antennas are used per sector, at spacing of ten or more wavelengths apart. This allows the operator to overcome the effects of fading due to physical phenomena such as multipath reception. Some amplification of the received signal as it leaves the antenna is often used to preserve the balance between uplink and downlink signal.

Base Station Controller

The base station controller (BSC) provides, classically, the *intelligence* behind the BTSs. Typically a BSC has tens or even hundreds of BTSs under its control. The BSC handles allocation of radio channels, receives measurements from the mobile phones, and controls handovers from BTS to BTS (except in the case of an inter-BSC handover in which case control is in part the responsibility of the anchor MSC). A key function of the BSC is to act as a concentrator where many different low capacity connections to BTSs (with relatively low utilisation) become reduced to a smaller number of connections towards the mobile switching center (MSC) (with a high level of utilisation). Overall, this means that networks are often structured to have many BSCs distributed into regions near their BTSs which are then connected to large centralised MSC sites.

The BSC is undoubtedly the most robust element in the BSS as it is not only a BTS controller but, for some vendors, a full switching center, as well as an SS7 node with connections to the MSC and serving GPRS support node (SGSN) (when using GPRS). It also provides all the required data to the operation support subsystem (OSS) as well as to the performance measuring centers.

A BSC is often based on a distributed computing architecture, with redundancy applied to critical functional units to ensure availability in the event of fault conditions. Redundancy often extends beyond the BSC equipment itself and is commonly used in the power supplies and in the transmission equipment providing the A-ter interface to PCU.

The databases for all the sites, including information such as carrier frequencies, frequency hopping lists, power reduction levels, receiving levels for cell border calculation, are stored in the BSC. This data is obtained directly from radio planning engineering which involves modelling of the signal propagation as well as traffic projections.

Transcoder

The transcoder is responsible for transcoding the voice channel coding between the coding used in the mobile network, and the coding used by the world's terrestrial circuit-switched network, the Public Switched Telephone Network. Specifically, GSM uses a regular pulse excited-long term prediction (RPE-LTP) coder for voice data between the mobile device and the BSS, but pulse code modulation (A-law or μ-law standardized in ITU G.711) upstream of the BSS. RPE-LPC coding results in a data rate for voice of 13 kbit/s where standard PCM coding results in 64 kbit/s. Because of this change in data rate *for the same voice call*, the transcoder also has a buffering function so that PCM 8-bit words can be recoded to construct GSM 20 ms traffic blocks.

Although transcoding (compressing/decompressing) functionality is defined as a base station function by the relevant standards, there are several vendors which have implemented the solution outside of the BSC. Some vendors have implemented it in a stand-alone rack using a proprietary interface. In Siemens' and Nokia's architecture, the transcoder is an identifiable separate sub-system which will normally be co-located with the MSC. In some of Ericsson's systems it is integrated to the MSC rather than the BSC. The reason for these designs is that if the compression of voice channels is done at the site of the MSC, the number of fixed transmission links between the BSS and MSC can be reduced, decreasing network infrastructure costs.

This subsystem is also referred to as the *transcoder and rate adaptation unit* (TRAU). Some networks use 32 kbit/s ADPCM on the terrestrial side of the network instead of 64 kbit/s PCM and the TRAU converts accordingly. When the traffic is not voice but data such as fax or email, the TRAU enables its rate adaptation unit function to give compatibility between the BSS and MSC data rates.

Packet Control Unit

The packet control unit (PCU) is a late addition to the GSM standard. It performs some of the processing tasks of the BSC, but for packet data. The allocation of channels between voice and data is controlled by the base station, but once a channel is allocated to the PCU, the PCU takes full control over that channel.

The PCU can be built into the base station, built into the BSC or even, in some proposed architectures, it can be at the SGSN site. In most of the cases, the PCU is a separate node communicating extensively with the BSC on the radio side and the SGSN on the Gb side.

BSS Interfaces

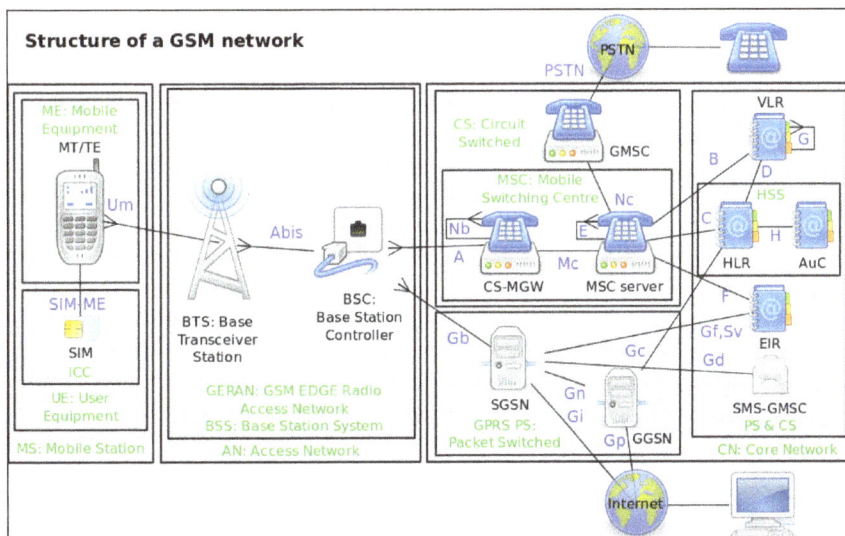

Image of the GSM network, showing the BSS interfaces to the MS, NSS and GPRS Core Network.

Um

The air interface between the mobile station (MS) and the BTS. This interface uses LAPDm protocol for signaling, to conduct call control, measurement reporting, handover, power control, authentication, authorization, location update and so on. Traffic and signaling are sent in bursts of 0.577 ms at intervals of 4.615 ms, to form data blocks each 20 ms.

Abis

The interface between the BTS and BSC. Generally carried by a DS-1, ES-1, or E1 TDM circuit. Uses TDM subchannels for traffic (TCH), LAPD protocol for BTS supervision and telecom signaling, and carries synchronization from the BSC to the BTS and MS.

A

The interface between the BSC and MSC. It is used for carrying traffic channels and the BSSAP user part of the SS7 stack. Although there are usually transcoding units between BSC and MSC, the signaling communication takes place between these two ending points and the transcoder unit doesn't touch the SS7 information, only the voice or CS data are transcoded or rate adapted.

Ater

The interface between the BSC and transcoder. It is a proprietary interface whose name depends on the vendor (for example Ater by Nokia), it carries the A interface information from the BSC leaving it untouched.

Gb

Connects the BSS to the SGSN in the GPRS core network.

Cellular Repeater

A cellular repeater (also known as cell phone signal booster or amplifier) is a type of bi-directional amplifier used to improve cell phone reception. A cellular repeater system commonly consists of a donor antenna that receives and transmits signal from nearby cell towers, coaxial cables, a signal amplifier, and an indoor rebroadcast antenna.

Donor Antenna

A "donor antenna" is typically installed by a window or on the roof a building and used to communicate back to a nearby cell tower. A donor antenna can be any of several types, but is usually directional or omnidirectional. An omnidirectional antenna (which broadcast in all directions) is typically used for a repeater system that amplify coverage for all cellular carriers. A directional antenna is used when a particular tower or carrier needs to be isolated for improvement. The use of a highly directional antenna can help improve the donor's signal-to-noise ratio, thus improving the quality of signal redistributed inside a building.

Indoor Antenna

Some cellular repeater systems can also include an omnidirectional antenna for rebroadcasting the signal indoors. Depending on attenuation from obstacles, the advantage of using an omnidirectional antenna is that the signal will be equally distributed in all directions.

Signal Amplifier

Cellular repeater systems include a signal amplifier. Standard GSM channel selective repeaters (operated by telecommunication operators for coverage of large areas and big buildings) have output power around 2 W, high power repeaters have output power around 10 W. The power gain is calculated by the following equation:

$$P_{dB} = 10\log_{10}\left(\frac{P}{P_0}\right).$$

A repeater needs to secure sufficient isolation between the donor and the service antenna. When the isolation is lower than actual gain plus a margin (of typically 5–15 dB), the repeater may go into in loop oscillation. This oscillation can cause interference to the cellular network.

The isolation may be improved by antenna type selection in a macro environment, which involves adjusting the angle between the donor and service antennas (ideally 180°), space separation (typically the vertical distance in the case of the tower installation between donor and service antenna is several meters), insertion into an attenuating environment (e.g. installing a metal mesh between donor and service antennas), and/or reduction of reflections (no near obstacles in front of the donor antenna such as trees or buildings).

Isolation can be also improved by integrated feature called ICE (interference cancellation equipment) offered in some products (e.g., NodeG, RFWindow). Activation of this feature has a negative impact on internal delay (higher delay => approximately +5 µs up to standard rep. delay) and consequently a shorter radius from donor site. Amplification and filtering introduce a delay (typically between 5 and 15 µs), depending on the type of repeater and features used. Additional distance also adds propagation delay.

Reasons for Weak Signal

Rural Areas

In many rural areas the housing density is too low to make construction of a new base station commercially viable. Installing a home cellular repeater may remedy this. In flat rural areas the signal is unlikely to suffer from multipath interference.

Building Construction Material

Certain construction materials can attenuate cell phone signal strength. Older buildings, such as churches, often block cellular signals. Any building that has a significant thickness of concrete, or a large amount of metal used in its construction, will attenuate the signal. Concrete floors are often poured onto a metal pan, which completely blocks most radio signals. Some solid foam insulation and some fiberglass insulation used in roofs or exterior walls have foil backing, which can reduce transmittance. Energy efficient windows and metal window screens are also very effective at blocking radio signals. Some materials have peaks in their absorption spectra, which decrease signal strength.

Building Size

Large buildings, such as warehouses, hospitals, and factories, often lack cellular reception. Low signal strength also tends to occur in underground areas (such as basements, and in shops and restaurants located towards the centre of shopping malls). In these cases, an external antenna is usually used.

Multipath Interference

Even in urban areas (which usually have strong cellular signals throughout), there may be dead zones caused by destructive interference of waves. These usually have an area of a few blocks and will usually only affect one of the two frequency ranges used by cell phones. This happens because different wavelengths of the different frequencies interfere destructively at different points. Directional antennas can be helpful at overcoming this issue since they may be used to select a single path from several

Diffraction and General Attenuation

The longer wavelengths have the advantage of diffracting more, and so line of sight is not as necessary to obtain a good signal. Because the frequencies that cell phones use are too high to reflect off the ionosphere as shortwave radio waves do, cell phone waves cannot travel via the ionosphere.

Different Operating Frequencies

Repeaters are available for all of the GSM frequency bands. Some repeaters will handle different types of networks (such as multi-mode GSM and UMTS). Repeater systems are available for certain Satellite phone systems, allowing these to be used indoors without a clear line of sight to the satellite.

GSM Network Architecture

A base station antenna carrying 2G GSM signals.

The GSM network architecture provided a simple and yet effective architecture to provide the services needed for a 2G cellular system. There were four main elements to the overall GSM network architecture and these could often be further split. Elements like the base station controller, MSC, AuC, HLR, VLR and the like are brought together to form the overall system.

The 2G GSM network architecture, although now superseded gives an excellent introduction into some of the basic capabilities required to set up a mobile phone network and how all the entities operate together.

GSM Network Architecture Elements

In order that the GSM system operates together as a complete system, the overall network architecture brings together a series of data network identities, each with several elements.

The GSM network architecture is defined in the GSM specifications and it can be grouped into four main areas:

- Network and Switching Subsystem (NSS).

- Base-Station Subsystem (BSS).

- Mobile station (MS).

- Operation and Support Subsystem (OSS).

The different elements of the GSM network operate together and the user is not aware of the different entities within the system.

As the GSM network is defined but he specifications and standards, it enables the system to operate reliably together regardless of the supplier of the different elements.

A basic diagram of the overall system architecture for 2G GSM with these four major elements is shown below:

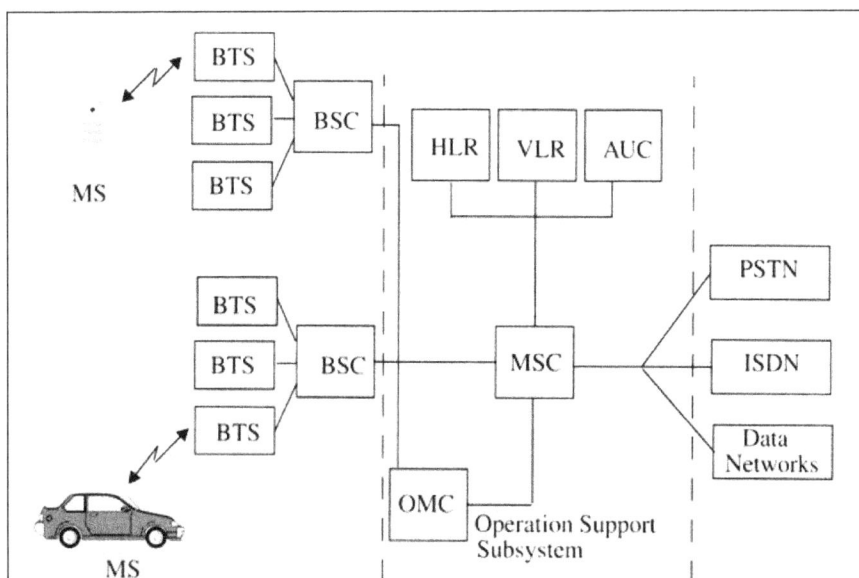

Simplified GSM Network Architecture Diagram.

Within this diagram the different network areas can be seen - they are grouped into the four areas that provide different functionality, but all operate to enable reliable mobile communications to be achieved.

The overall network architecture provided to be very successful and was developed further to enable 2G evolution to carry data and then with further evolutions to allow 3G to be established.

Network Switching Subsystem (NSS)

The GSM system architecture contains a variety of different elements, and is often termed the core network. It is essentially a data network with a various entities that provide the main control and interfacing for the whole mobile network. The major elements within the core network include:

- Mobile Services Switching Centre (MSC): The main element within the core network area of the overall GSM network architecture is the Mobile switching Services Centre (MSC). The MSC acts like a normal switching node within a PSTN or ISDN, but also provides additional functionality to enable the requirements of a mobile user to be supported. These include registration, authentication, call location, inter-MSC handovers and call routing to a mobile subscriber. It also provides an interface to the PSTN so that calls can be routed from the mobile network to a phone connected to a landline. Interfaces to other MSCs are provided to enable calls to be made to mobiles on different networks.

- Home Location Register (HLR): This database contains all the administrative information about each subscriber along with their last known location. In this way, the GSM network is able to route calls to the relevant base station for the MS. When a user switches on their phone, the phone registers with the network and from this it is possible to determine which BTS it communicates with so that incoming calls can be routed appropriately. Even when the phone is not active (but switched on) it re-registers periodically to ensure that the network (HLR) is aware of its latest position. There is one HLR per network, although it may be distributed across various sub-centres to for operational reasons.

- Visitor Location Register (VLR): This contains selected information from the HLR that enables the selected services for the individual subscriber to be provided. The VLR can be implemented as a separate entity, but it is commonly realised as an integral part of the MSC, rather than a separate entity. In this way access is made faster and more convenient.

- Equipment Identity Register (EIR): The EIR is the entity that decides whether given mobile equipment may be allowed onto the network. Each mobile equipment has a number known as the International Mobile Equipment Identity. This number is installed in the equipment and is checked by the network during registration. Dependent upon the information held in the EIR, the mobile may be allocated one of three states - allowed onto the network, barred access, or monitored in case its problems.

- Authentication Centre (AuC): The AuC is a protected database that contains the secret key also contained in the user's SIM card. It is used for authentication and for ciphering on the radio channel.

- Gateway Mobile Switching Centre (GMSC): The GMSC is the point to which a ME terminating call is initially routed, without any knowledge of the MS's location. The GMSC is thus in charge of obtaining the MSRN (Mobile Station Roaming Number) from the HLR based on the MSISDN (Mobile Station ISDN number, the "directory number" of a MS) and routing the call to the correct visited MSC. The "MSC" part of the term GMSC is misleading, since the gateway operation does not require any linking to an MSC.

- SMS Gateway (SMS-G): The SMS-G or SMS gateway is the term that is used to collectively describe the two Short Message Services Gateways defined in the GSM standards. The two gateways handle messages directed in different directions. The SMS-GMSC (Short Message Service Gateway Mobile Switching Centre) is for short messages being sent to an ME. The SMS-IWMSC (Short Message Service Inter-Working Mobile Switching Centre) is used for short messages originated with a mobile on that network. The SMS-GMSC role is similar to that of the GMSC, whereas the SMS-IWMSC provides a fixed access point to the Short Message Service Centre.

These entities were the main ones used within the GSM network. They were typically co-located, but often the overall core network was distributed around the country where the network was located. This gave some resilience in case of failure.

Although the GSM system was essential a voice system, the core network was a data network as all signals were handled digitally.

Base Station Subsystem (BSS)

The Base Station Subsystem (BSS) section of the 2G GSM network architecture that is fundamentally associated with communicating with the mobiles on the network.

It consists of two elements:

- Base Transceiver Station (BTS): The BTS used in a GSM network comprises the radio transmitter receivers, and their associated antennas that transmit and receive to directly communicate with the mobiles. The BTS is the defining element for each cell. The BTS communicates with the mobiles and the interface between the two is known as the Um interface with its associated protocols.

- Base Station Controller (BSC): The BSC forms the next stage back into the GSM network. It controls a group of BTSs, and is often co-located with one of the BTSs in its group. It manages the radio resources and controls items such as handover within the group of BTSs, allocates channels and the like. It communicates with the BTSs over what is termed the Abis interface.

The base station subsystem element of the GSM network utilised the radio access technology to enable a number of users to access the system concurrently. Each channel supported up to eight users and by enabling a base station to have several channels, a large number of subscribers could be accommodated by each base station.

Base stations are carefully located by the network provider to enable complete coverage of an area. The area being covered bay a base station often being referred to as a cell.

As it is not possible to prevent overlap of the signals into the adjacent cells, channels used in one cell are not used in the next. In this way interference which would reduce call quality is reduced whilst still maintaining sufficient frequency re-use. It is important to have the different BTSs linked with the BSS and the BSSs linked back to the core network.

A variety of technologies were used to achieve this. As data rates used within he GSM network were relatively low, E1 or T1 lines were often used, especially for linking the BSS back to the core network. As more data was required with increasing usage of the GSM network, and also as other cellular technologies like 3G became more widespread, many links used carrier grade Ethernet.

Often remote BTSs were linked using small microwave links as this could reduce the need for the installation of specific lines if none were available. As base stations often needed to be located to provide good coverage rather than in areas where lines could be installed, the microwave link option provided an attractive method for providing a data link for the network.

Mobile Station

Mobile stations (MS), mobile equipment (ME) or as they are most widely known, cell or mobile phones are the section of a GSM cellular network that the user sees and operates. In recent years their size has fallen dramatically while the level of functionality has greatly increased. A further advantage is that the time between charges has significantly increased.

There are a number of elements to the cell phone, although the two main elements are the main hardware and the SIM.

The hardware itself contains the main elements of the mobile phone including the display, case, battery, and the electronics used to generate the signal, and process the data receiver and to be transmitted.

The mobile station, or ME also contains a number known as the International Mobile Equipment Identity (IMEI). This is installed in the phone at manufacture and "cannot" be changed. It is accessed by the network during registration to check whether the equipment has been reported as stolen.

The SIM or Subscriber Identity Module contains the information that provides the identity of the user to the network. It contains are variety of information including a number known as the International Mobile Subscriber Identity (IMSI). As this is included in the SIM, and it means that by moving the SIM card from one mobile to another, the user could easily change mobiles. The ease of changing mobiles whilst keeping the same number meant that people would regularly upgrade, thereby creating a further revenue stream for network providers and helping to increase the overall financial success of GSM.

Operation and Support Subsystem (OSS)

The OSS or operation support subsystem is an element within the overall GSM network architecture that is connected to components of the NSS and the BSC. It is used to control and monitor the overall GSM network and it is also used to control the traffic load of the BSS. It must be noted that as the number of BS increases with the scaling of the subscriber population some of the maintenance tasks are transferred to the BTS, allowing savings in the cost of ownership of the system.

The 2G GSM network architecture follows a logical method of operation. It is far simpler than current mobile phone network architectures which use software defined entities to enable very flexible operation. However the 2G GSM architecture does show the voice and operational basic functions that are needed and how they fit together. As the GSM system was all digital, the network was a data network.

Network Switching Subsystem

Network switching subsystem (NSS) (or GSM core network) is the component of a GSM system that carries out call out and mobility management functions for mobile phones roaming on the network of base stations. It is owned and deployed by mobile phone operators and allows mobile devices to communicate with each other and telephones in the wider public switched telephone network (PSTN). The architecture contains specific features and functions which are needed because the phones are not fixed in one location.

The NSS originally consisted of the circuit-switched core network, used for traditional GSM services such as voice calls, SMS, and circuit switched data calls. It was extended with an overlay architecture to provide packet-switched data services known as the GPRS core network. This allows mobile phones to have access to services such as WAP, MMS and the Internet.

Mobile Switching Center (MSC)

The JNM mobile switching center (MSC) is the primary service delivery node for GSM/CDMA, responsible for routing voice calls and SMS as well as other services (such as conference calls, FAX and circuit switched data).

The MSC sets up and releases the end-to-end connection, handles mobility and hand-over requirements during the call and takes care of charging and real time prepaid account monitoring.

In the GSM mobile phone system, in contrast with earlier analogue services, fax and data information is sent digitally encoded directly to the MSC. Only at the MSC is this re-coded into an "analogue" signal (although actually this will almost certainly mean sound is encoded digitally as a pulse-code modulation (PCM) signal in a 64-kbit/s timeslot, known as a DS0 in America).

There are various different names for MSCs in different contexts which reflects their complex role in the network, all of these terms though could refer to the same MSC, but doing different things at different times.

The gateway MSC (G-MSC) is the MSC that determines which "visited MSC" (V-MSC) the subscriber who is being called is currently located at. It also interfaces with the PSTN. All mobile to mobile calls and PSTN to mobile calls are routed through a G-MSC. The term is only valid in the context of one call, since any MSC may provide both the gateway function and the visited MSC function. However, some manufacturers design dedicated high capacity MSCs which do not have any base station subsystems (BSS) connected to them. These MSCs will then be the gateway MSC for many of the calls they handle.

The visited MSC (V-MSC) is the MSC where a customer is currently located. The visitor location register (VLR) associated with this MSC will have the subscriber's data in it.

The anchor MSC is the MSC from which a handover has been initiated. The target MSC is the MSC toward which a handover should take place. A mobile switching center server is a part of the redesigned MSC concept starting from 3GPP Release 4.

Mobile Switching Center Server (MSC-Server, MSCS or MSS)

The mobile switching center server is a soft-switch variant (therefore it may be referred to as mobile soft switch, MSS) of the mobile switching center, which provides circuit-switched calling mobility management, and GSM services to the mobile phones roaming within the area that it serves. Functionality enables split control between (signaling) and user plane (bearer in network element called as media gateway/MG), which guarantees better placement of network elements within the network.

MSS and media gateway (MGW) makes it possible to cross-connect circuit switched calls switched by using IP, ATM AAL2 as well as TDM.

The term Circuit switching (CS) used here originates from traditional telecommunications systems. However, modern MSS and MGW devices mostly use generic Internet technologies and form next-generation telecommunication networks. MSS software may run on generic computers or virtual machines in cloud environment.

Other GSM Core Network Elements Connected to the MSC

The MSC connects to the following elements:

- The home location registering (HLR) for obtaining data about the SIM and mobile services ISDN number (MSISDN; i.e., the telephone number).

- The base station subsystems (BSS) which handles the radio communication with 2G and 2.5G mobile phones.

- The UMTS terrestrial radio access network (UTRAN) which handles the radio communication with 3G mobile phones.

- The visitor location register (VLR) provides subscriber information when the subscriber is outside its home network.

- Other MSCs for procedures such as hand over.

Procedures Implemented

Tasks of the MSC include:

- Delivering calls to subscribers as they arrive based on information from the VLR.

- Connecting outgoing calls to other mobile subscribers or the PSTN.

- Delivering SMSs from subscribers to the short message service center (SMSC) and vice versa.

- Arranging handovers from BSC to BSC.

- Carrying out handovers from this MSC to another.

- Supporting supplementary services such as conference calls or call hold.

- Generating billing information.

Home Location Register (HLR)

The home location register (HLR) is a central database that contains details of each mobile phone subscriber that is authorized to use the GSM core network. There can be several logical, and physical, HLRs per public land mobile network (PLMN), though one international mobile subscriber identity (IMSI)/MSISDN pair can be associated with only one logical HLR (which can span several physical nodes) at a time.

The HLRs store details of every SIM card issued by the mobile phone operator. Each SIM has a unique identifier called an IMSI which is the primary key to each HLR record.

Another important item of data associated with the SIM are the MSISDNs, which are the telephone numbers used by mobile phones to make and receive calls. The primary MSISDN is the number used for making and receiving voice calls and SMS, but it is possible for a SIM to have other secondary MSISDNs associated with it for fax and data calls. Each MSISDN is also a unique key to the HLR record. The HLR data is stored for as long as a subscriber remains with the mobile phone operator.

Examples of other data stored in the HLR against an IMSI record is:

- GSM services that the subscriber has requested or been given.

- General Packet Radio Service (GPRS) settings to allow the subscriber to access packet services.

- Current location of subscriber (VLR and serving GPRS support node/SGSN).

- Call divert settings applicable for each associated MSISDN.

The HLR is a system which directly receives and processes MAP transactions and messages from elements in the GSM network, for example, the location update messages received as mobile phones roam around.

Other GSM Core Network Elements Connected to the HLR

The HLR connects to the following elements:

- The G-MSC for handling incoming calls.

- The VLR for handling requests from mobile phones to attach to the network.

- The SMSC for handling incoming SMSs.

- The voice mail system for delivering notifications to the mobile phone that a message is waiting.

- The AuC for authentication and ciphering and exchange of data (triplets).

Procedures Implemented

The main function of the HLR is to manage the fact that SIMs and phones move around a lot. The following procedures are implemented to deal with this:

- Manage the mobility of subscribers by means of updating their position in administrative areas called 'location areas', which are identified with a LAC. The action of a user of moving from one LA to another is followed by the HLR with a Location area update procedure.

- Send the subscriber data to a VLR or SGSN when a subscriber first roams there.

- Broker between the G-MSC or SMSC and the subscriber's current VLR in order to allow incoming calls or text messages to be delivered.

- Remove subscriber data from the previous VLR when a subscriber has roamed away from it.

- Responsible for all SRI related queries (i.e. for invoke SRI, HLR should give sack SRI or SRI reply).

Authentication Center (AuC)

The authentication center (AuC) is a function to authenticate each SIM card that attempts to connect to the gsm core network (typically when the phone is powered on). Once the authentication is successful, the HLR is allowed to manage the SIM and services described above. An encryption key is also generated that is subsequently used to encrypt all wireless communications (voice, SMS, etc.) between the mobile phone and the GSM core network.

If the authentication fails, then no services are possible from that particular combination of SIM card and mobile phone operator attempted. There is an additional form of identification check performed on the serial number of the mobile phone but this is not relevant to the AuC processing.

Proper implementation of security in and around the AuC is a key part of an operator's strategy to avoid SIM cloning.

The AuC does not engage directly in the authentication process, but instead generates data known as *triplets* for the MSC to use during the procedure. The security of the process depends upon a shared secret between the AuC and the SIM called the K_i. The K_i is securely burned into the SIM during manufacture and is also securely replicated onto the AuC. This K_i is never transmitted between the AuC and SIM, but is combined with the IMSI to produce a challenge/response for identification purposes and an encryption key called K_c for use in over the air communications.

Other GSM Core Network Elements Connected to the AuC

The AuC connects to the following elements:

- The MSC which requests a new batch of triplet data for an IMSI after the previous data have been used. This ensures that same keys and challenge responses are not used twice for a particular mobile.

Procedures Implemented

The AuC stores the following data for each IMSI:

- The Ki,

- Algorithm id. (the standard algorithms are called A3 or A8, but an operator may choose a proprietary one).

When the MSC asks the AuC for a new set of triplets for a particular IMSI, the AuC first generates a random number known as *RAND*. This *RAND* is then combined with the K_i to produce two numbers as follows:

- The Ki and RAND are fed into the A3 algorithm and the signed response (SRES) is calculated,

- The Ki and RAND are fed into the A8 algorithm and a session key called Kc is calculated.

The numbers (*RAND*, SRES, K_c) form the triplet sent back to the MSC. When a particular IMSI requests access to the GSM core network, the MSC sends the *RAND* part of the triplet to the SIM. The SIM then feeds this number and the K_i (which is burned onto the SIM) into the A3 algorithm as appropriate and an SRES is calculated and sent back to the MSC. If this SRES matches with the SRES in the triplet (which it should if it is a valid SIM), then the mobile is allowed to attach and proceed with GSM services.

After successful authentication, the MSC sends the encryption key K_c to the base station controller (BSC) so that all communications can be encrypted and decrypted. Of course, the mobile phone can generate the K_c itself by feeding the same RAND supplied during authentication and the K_i into the A8 algorithm.

The AuC is usually collocated with the HLR, although this is not necessary. Whilst the procedure is secure for most everyday use, it is by no means hack proof. Therefore, a new set of security methods was designed for 3G phones.

A3 Algorithm is used to encrypt Global System for Mobile Communications (GSM) cellular communications. In practice, A3 and A8 algorithms are generally implemented together (known as A3/A8). An A3/A8 algorithm is implemented in Subscriber Identity Module (SIM) cards and in GSM network Authentication Centers. It is used to authenticate the customer and generate a key for encrypting voice and data traffic, as defined in 3GPP TS 43.020 (03.20 before Rel-4). Development of A3 and A8 algorithms is considered a matter for individual GSM network operators, although example implementations are available.

Visitor Location Register (VLR)

The Visitor Location Register (VLR) is a database of the MSs (Mobile stations) that have roamed into the jurisdiction of the MSC (Mobile Switching Center) which it serves. Each main base station in the network is served by exactly one VLR (one BTS may be served by many MSCs in case of MSC in pool), hence a subscriber cannot be present in more than one VLR at a time.

The data stored in the VLR has either been received from the Home Location Register (HLR), or collected from the MS. In practice, for performance reasons, most vendors integrate the VLR directly to the V-MSC and, where this is not done, the VLR is very tightly linked with the MSC via a proprietary interface. Whenever an MSC detects a new MS in its network, in addition to creating a new record in the VLR, it also updates the HLR of the mobile subscriber, apprising it of the new location of that MS. If VLR data is corrupted it can lead to serious issues with text messaging and call services.

Data Stored Include

- IMSI (the subscriber's identity number).

- Authentication data.

- MSISDN (the subscriber's phone number).

- GSM services that the subscriber is allowed to access.

- Access point (GPRS) subscribed.

- The HLR address of the subscriber.

- SCP Address(For Prepaid Subscriber).

Procedures Implemented

The primary functions of the VLR are:

- To inform the HLR that a subscriber has arrived in the particular area covered by the VLR.

- To track where the subscriber is within the VLR area (location area) when no call is on-going.

- To allow or disallow which services the subscriber may use.

- To allocate roaming numbers during the processing of incoming calls.

- To purge the subscriber record if a subscriber becomes inactive whilst in the area of a VLR. The VLR deletes the subscriber's data after a fixed time period of inactivity and informs the HLR (e.g., when the phone has been switched off and left off or when the subscriber has moved to an area with no coverage for a long time).

- To delete the subscriber record when a subscriber explicitly moves to another, as instructed by the HLR.

Equipment Identity Register (EIR)

The equipment identity register is often integrated to the HLR. The EIR keeps a list of mobile phones (identified by their IMEI) which are to be banned from the network or monitored. This is designed to allow tracking of stolen mobile phones. In theory all data about all stolen mobile phones should be distributed to all EIRs in the world through a Central EIR. It is clear, however, that there are some countries where this is not in operation. The EIR data does not have to change in real time, which means that this function can be less distributed than the function of the HLR. The EIR is a database that contains information about the identity of the mobile equipment that prevents calls from stolen, unauthorized or defective mobile stations. Some EIR also have the capability to log Handset attempts and store it in a log file.

Other Support Functions

Connected more or less directly to the GSM core network are many other functions.

Billing Center (BC)

The billing center is responsible for processing the toll tickets generated by the VLRs and HLRs and generating a bill for each subscriber. It is also responsible for generating billing data of roaming subscriber.

Multimedia Messaging Service Center (MMSC)

The multimedia messaging service center supports the sending of multimedia messages (e.g., images, audio, video and their combinations) to (or from) MMS-bluetooth.

Voicemail System (VMS)

The voicemail system records and stores voicemail. which may have to pay.

Lawful Interception Functions

According to U.S. law, which has also been copied into many other countries, especially in Europe, all telecommunications equipment must provide facilities for monitoring the calls of selected users. There must be some level of support for this built into any of the different elements. The concept of *lawful interception* is also known, following the relevant U.S. law, as CALEA. Generally, lawful Interception implementation is similar to the implementation of conference call. While A and B are talking with each other, C can join the call and listen silently.

GPRS Architecture

GPRS architecture works on the same procedure like GSM network, but, has additional entities that allow packet data transmission. This data network overlaps a second-generation GSM network providing packet data transport at the rates from 9.6 to 171 kbps. Along with the packet data transport the GSM network accommodates multiple users to share the same air interface resources concurrently.

Following is the GPRS Architecture diagram:

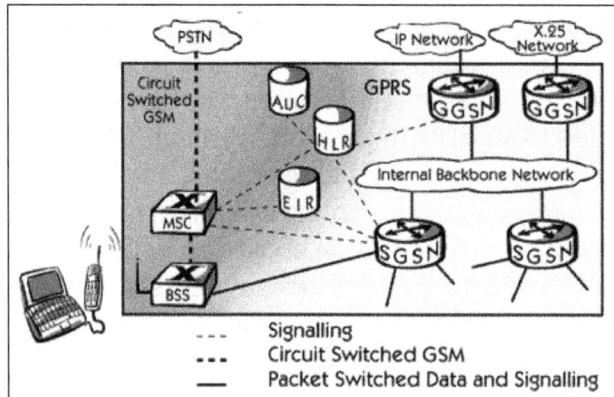

GPRS attempts to reuse the existing GSM network elements as much as possible, but to effectively build a packet-based mobile cellular network, some new network elements, interfaces, and protocols for handling packet traffic are required.

Therefore, GPRS requires modifications to numerous GSM network elements as summarized below:

GSM Network Element	Modification or Upgrade Required for GPRS.
Mobile Station (MS)	New Mobile Station is required to access GPRS services. These new terminals will be backward compatible with GSM for voice calls.
BTS	A software upgrade is required in the existing Base Transceiver Station(BTS).
BSC	The Base Station Controller (BSC) requires a software upgrade and the installation of new hardware called the packet control unit (PCU). The PCU directs the data traffic to the GPRS network and can be a separate hardware element associated with the BSC.
GPRS Support Nodes (GSNs)	The deployment of GPRS requires the installation of new core network elements called the serving GPRS support node (SGSN) and gateway GPRS support node (GGSN).
Databases (HLR, VLR, etc.)	All the databases involved in the network will require software upgrades to handle the new call models and functions introduced by GPRS.

GPRS Mobile Stations

New Mobile Stations (MS) are required to use GPRS services because existing GSM phones do not handle the enhanced air interface or packet data. A variety of MS can exist, including a high-speed version of current phones to support high-speed data access, a new PDA device with an embedded GSM phone, and PC cards for laptop computers. These mobile stations are backward compatible for making voice calls using GSM.

GPRS Base Station Subsystem

Each BSC requires the installation of one or more Packet Control Units (PCUs) and a software upgrade. The PCU provides a physical and logical data interface to the Base Station Subsystem (BSS) for packet data traffic. The BTS can also require a software upgrade but typically does not require hardware enhancements.

When either voice or data traffic is originated at the subscriber mobile, it is transported over the air interface to the BTS, and from the BTS to the BSC in the same way as a standard GSM call. However, at the output of the BSC, the traffic is separated; voice is sent to the Mobile Switching Center (MSC) per standard GSM, and data is sent to a new device called the SGSN via the PCU over a Frame Relay interface.

GPRS Support Nodes

Following two new components, called Gateway GPRS Support Nodes (GSNs) and, Serving GPRS Support Node (SGSN) are added:

Gateway GPRS Support Node (GGSN)

The Gateway GPRS Support Node acts as an interface and a router to external networks. It contains routing information for GPRS mobiles, which is used to tunnel packets through the IP based internal backbone to the correct Serving GPRS Support Node. The GGSN also collects charging information connected to the use of the external data networks and can act as a packet filter for incoming traffic.

Serving GPRS Support Node (SGSN)

The Serving GPRS Support Node is responsible for authentication of GPRS mobiles, registration of mobiles in the network, mobility management, and collecting information on charging for the use of the air interface.

Internal Backbone

The internal backbone is an IP based network used to carry packets between different GSNs. Tunnelling is used between SGSNs and GGSNs, so the internal backbone does not need any information about domains outside the GPRS network. Signalling from a GSN to a MSC, HLR or EIR is done using SS7.

Routing Area

GPRS introduces the concept of a Routing Area. This concept is similar to Location Area in GSM, except that it generally contains fewer cells. Because routing areas are smaller than location areas, less radio resources are used while broadcasting a page message.

LTE Network Architecture

Long Term Evolution (LTE) has been designed to support only packet-switched services. It aims to provide seamless Internet Protocol (IP) connectivity between user equipment (UE) and the packet data network (PDN), without any disruption to the end users' applications during mobility.

While the term "LTE" encompasses the evolution of the Universal Mobile Telecommunications System (UMTS) radio access through the Evolved UTRAN (E-UTRAN), it is accompanied by an evolution of the non-radio aspects under the term "System Architecture Evolution" (SAE), which includes the Evolved Packet Core (EPC) network. Together LTE and SAE comprise the Evolved Packet System (EPS).

EPS uses the concept of EPS bearers to route IP traffic from a gateway in the PDN to the UE. A bearer is an IP packet flow with a defined quality of service (QoS) between the gateway and the UE. The E-UTRAN and EPC together set up and release bearers as required by applications.

EPS provides the user with IP connectivity to a PDN for accessing the Internet, as well as for running services such as Voice over IP (VoIP). An EPS bearer is typically associated with a QoS. Multiple bearers can be established for a user in order to provide different QoS streams or connectivity to different PDNs. For example, a user might be engaged in a voice (VoIP) call while at the same time performing web browsing or FTP download. A VoIP bearer would provide the necessary QoS for the voice call, while a best-effort bearer would be suitable for the web browsing or FTP session.

The network must also provide sufficient security and privacy for the user and protection for the network against fraudulent use.

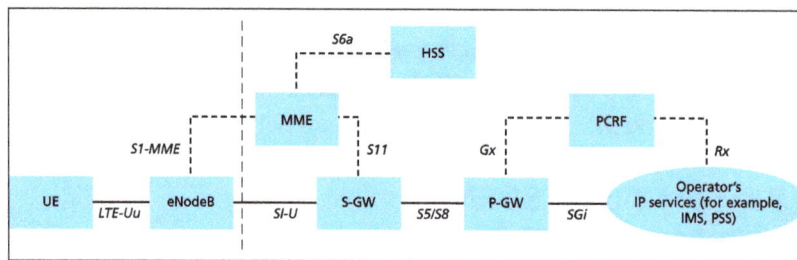

The EPS network elements.

This is achieved by means of several EPS network elements that have different roles. Figure shows the overall network architecture, including the network elements and the standardized interfaces. At a high level, the network is comprised of the CN (EPC) and the access network E-UTRAN. While the CN consists of many logical nodes, the access network is made up of essentially just one node, the evolved NodeB (eNodeB), which connects to the UEs. Each of these network elements is interconnected by means of interfaces that are standardized in order to allow multi-vendor interoperability. This gives network operators the possibility to source different network elements from different vendors. In fact, network operators may choose in their physical implementations to split or merge these logical network elements depending on commercial considerations. The functional split between the EPC and E-UTRAN is shown in figure.

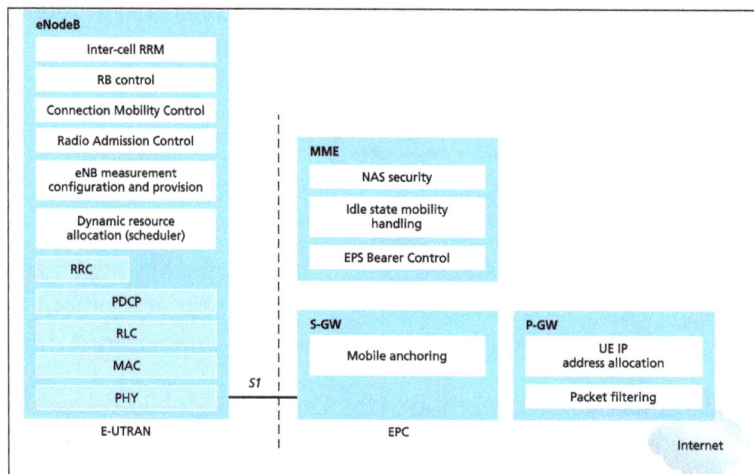

Functional split between E-UTRAN and EPC.

The Core Network

The core network (called EPC in SAE) is responsible for the overall control of the UE and establishment of the bearers. The main logical nodes of the EPC are:

- PDN Gateway (P-GW).

- Serving Gateway (S-GW).

- Mobility Management Entity (MME).

In addition to these nodes, EPC also includes other logical nodes and functions such as the Home Subscriber Server (HSS) and the Policy Control and Charging Rules Function (PCRF). Since the EPS only provides a bearer path of a certain QoS, control of multimedia applications such as VoIP is provided by the IP Multimedia Subsystem (IMS), which is considered to be outside the EPS itself.

The logical CN nodes are shown in Figure and discussed in more detail below:

- PCRF: The Policy Control and Charging Rules Function are responsible for policy control decision-making, as well as for controlling the flow-based charging functionalities in the Policy Control Enforcement Function (PCEF), which resides in the P-GW. The PCRF provides the QoS authorization (QoS class identifier [QCI] and bit rates) that decides how a certain data flow will be treated in the PCEF and ensures that this is in accordance with the user's subscription profile.

- HSS: The Home Subscriber Server contains users' SAE subscription data such as the EPS-subscribed QoS profile and any access restrictions for roaming. It also holds information about the PDNs to which the user can connect. This could be in the form of an access point name (APN) (which is a label according to DNS naming conventions describing the access point to the PDN) or a PDN address (indicating subscribed IP address(es)). In addition the HSS holds dynamic information such as the identity of the MME to which the user is currently attached or registered. The HSS may also integrate the authentication center (AUC), which generates the vectors for authentication and security keys.

- P-GW: The PDN Gateway is responsible for IP address allocation for the UE, as well as QoS enforcement and flow-based charging according to rules from the PCRF. It is responsible for the filtering of downlink user IP packets into the different QoS-based bearers. This is performed based on Traffic Flow Templates (TFTs). The P-GW performs QoS enforcement for guaranteed bit rate (GBR) bearers. It also serves as the mobility anchor for interworking with non-3GPP technologies such as CDMA2000 and WiMAX networks.

- S-GW: All user IP packets are transferred through the Serving Gateway, which serves as the local mobility anchor for the data bearers when the UE moves between eNodeBs. It also retains the information about the bearers when the UE is in the idle state (known as "EPS Connection Management — IDLE" [ECM-IDLE]) and temporarily buffers downlink data while the MME initiates paging of the UE to reestablish the bearers. In addition, the S-GW performs some administrative functions in the visited network such as collecting information for charging (for example, the volume of data sent to or received from the user) and lawful interception. It also serves as the mobility anchor for interworking with other 3GPP technologies such as general packet radio service (GPRS) and UMTS.

- MME: The Mobility Management Entity (MME) is the control node that processes the signaling between the UE and the CN. The protocols running between the UE and the CN are known as the Non Access Stratum (NAS) protocols.

The main functions supported by the MME can be classified as:

- Functions related to bearer management: This includes the establishment, maintenance and release of the bearers and is handled by the session management layer in the NAS protocol.

- Functions related to connection management: This includes the establishment of the connection and security between the network and UE and is handled by the connection or mobility management layer in the NAS protocol layer.

Non Access Stratum Procedures

The Non Access Stratum procedures, especially the connection management procedures, are fundamentally similar to UMTS. The main change from UMTS is that EPS allows concatenation of some procedures to allow faster establishment of the connection and the bearers.

The MME creates a UE context when a UE is turned on and attaches to the network. It assigns a unique short temporary identity termed the SAE Temporary Mobile Subscriber Identity (S-TM-SI) to the UE that identifies the UE context in the MME. This UE context holds user subscription information downloaded from the HSS. The local storage of subscription data in the MME allows faster execution of procedures such as bearer establishment since it removes the need to consult the HSS every time. In addition, the UE context also holds dynamic information such as the list of bearers that are established and the terminal capabilities.

To reduce the overhead in the E-UTRAN and processing in the UE, all UE-related information in the access network, including the radio bearers, can be released during long periods of data inactivity. This is the ECM-IDLE state. The MME retains the UE context and the information about the established bearers during these idle periods.

To allow the network to contact an ECM-IDLE UE, the UE updates the network as to its new location whenever it moves out of its current tracking area (TA); this procedure is called a tracking area update. The MME is responsible for keeping track of the user location while the UE is in ECM-IDLE.

When there is a need to deliver downlink data to an ECM-IDLE UE, the MME sends a paging message to all the eNodeBs in its current TA, and the eNodeBs page the UE over the radio interface. On receipt of a paging message, the UE performs a Service Request procedure, which results in moving the UE to the ECM-CONNECTED state. UE-related information is thereby created in the E-UTRAN, and the radio bearers are reestablished. The MME is responsible for the reestablishment of the radio bearers and updating the UE context in the eNodeB. This transition between the UE states is called an idle-to-active transition. To speed up the idle-to-active transition and bearer establishment, EPS supports concatenation of the NAS and Access Stratum (AS) procedures for bearer activation. Some interrelationship between the NAS and AS protocols is intentionally used to allow procedures to run simultaneously rather than sequentially, as in UMTS. For example, the bearer establishment procedure can be executed by the network without waiting for the completion of the security procedure.

Security functions are the responsibility of the MME for both signaling and user data. When a UE attaches with the network, a mutual authentication of the UE and the network is performed between the UE and the MME/HSS. This authentication function also establishes the security keys that are used for encryption of the bearers.

The Access Network

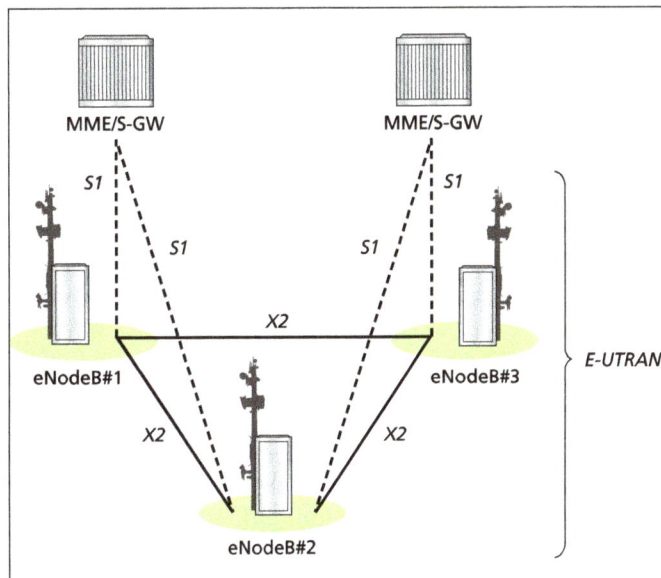

Overall E-UTRAN architecture.

The access network of LTE, E-UTRAN, simply consists of a network of eNodeBs, as illustrated in Figure. For normal user traffic (as opposed to broadcast), there is no centralized controller in E-UTRAN; hence the E-UTRAN architecture is said to be flat.

The eNodeBs are normally interconnected with each other by means of an interface known as "X2" and to the EPC by means of the S1 interface — more specifically, to the MME by means of the S1-MME interface and to the S-GW by means of the S1-U interface.

The protocols that run between the eNodeBs and the UE are known as the "AS protocols."

The E-UTRAN is responsible for all radio-related functions, which can be summarized briefly as:

- Radio Resource Management (RRM): This covers all functions related to the radio bearers, such as radio bearer control, radio admission control, radio mobility control, scheduling and dynamic allocation of resources to UEs in both uplink and downlink.

- Header Compression: This helps to ensure efficient use of the radio interface by compressing the IP packet headers that could otherwise represent a significant overhead, especially for small packets such as VoIP.

- Security: All data sent over the radio interface is encrypted.

- Connectivity to the EPC: This consists of the signaling toward MME and the bearer path toward the S-GW.

On the network side, all of these functions reside in the eNodeBs, each of which can be responsible for managing multiple cells. Unlike some of the previous second- and third-generation technologies, LTE integrates the radio controller function into the eNodeB. This allows tight interaction between the different protocol layers of the radio access network (RAN), thus reducing latency and improving efficiency. Such distributed control eliminates the need for a high-availability, processing-intensive controller, which in turn has the potential to reduce costs and avoid "single points of failure." Furthermore, as LTE does not support soft handover there is no need for a centralized data-combining function in the network.

One consequence of the lack of a centralized controller node is that, as the UE moves, the network must transfer all information related to a UE, that is, the UE context, together with any buffered data, from one eNodeB to another. Mechanisms are therefore needed to avoid data loss during handover.

An important feature of the S1 interface linking the access network to the CN is known as "S1-flex." This is a concept whereby multiple CN nodes (MME/S-GWs) can serve a common geographical area, being connected by a mesh network to the set of eNodeBs in that area. An eNodeB may thus be served by multiple MME/S-GWs, as is the case for eNodeB #2 in figure. The set of MME/S-GW nodes that serves a common area is called an MME/S-GW pool, and the area covered by such a pool of MME/S-GWs is called a pool area. This concept allows UEs in the cell or cells controlled by one eNodeB to be shared between multiple CN nodes, thereby providing a possibility for load sharing and also eliminating single points of failure for the CN nodes. The UE context normally remains with the same MME as long as the UE is located within the pool area.

Roaming Architecture

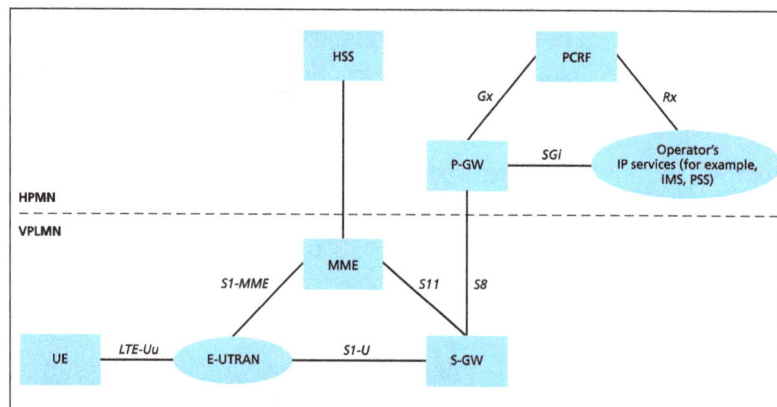

Roaming architecture for 3GPP accesses with P-GW in home network.

A network run by one operator in one country is known as a "public land mobile network (PLMN)." Roaming, where users are allowed to connect to PLMNs other than those to which they are directly subscribed, is a powerful feature for mobile networks, and LTE/SAE is no exception. A roaming user is connected to the E-UTRAN, MME and S-GW of the visited LTE network. However, LTE/SAE allows the P-GW of either the visited or the home network to be used, as shown in Figure. Using the home network's P-GW allows the user to access the home operator's services even while in a visited network. A P-GW in the visited network allows a "local breakout" to the Internet in the visited network.

Interworking with other Networks

EPS also supports interworking and mobility (handover) with networks using other Radio Access Technologies (RATs), notably Global System for Mobile Communications (GSM), UMTS, CDMA2000 and WiMAX. The architecture for interworking with 2G and 3G GPRS/UMTS networks is shown in figure. The S-GW acts as the mobility anchor for interworking with other 3GPP technologies such as GSM and UMTS, while the P-GW serves as an anchor allowing seamless mobility to non-3GPP networks such as CDMA2000 or WiMAX. The P-GW may also support a Proxy Mobile Internet Protocol (PMIP)-based interface.

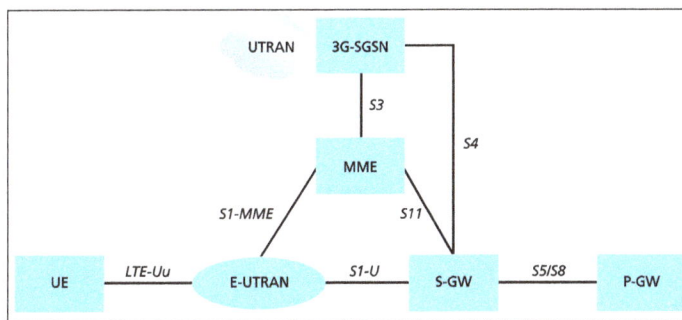

Architecture for 3G UMTS interworking.

Protocol Architecture

We outline here the radio protocol architecture of E-UTRAN.

User Plane

An IP packet for a UE is encapsulated in an EPC-specific protocol and tunneled between the P-GW and the eNodeB for transmission to the UE. Different tunneling protocols are used across different interfaces. A 3GPP-specific tunneling protocol called the GPRS Tunneling Protocol (GTP) is used over the CN interfaces, S1 and S5/S8.

The E-UTRAN user plane protocol stack is shown in blue in Figure, consisting of the Packet Data Convergence Protocol (PDCP), Radio Link Control (RLC) and Medium Access Control (MAC) sublayers that are terminated in the eNodeB on the network side.

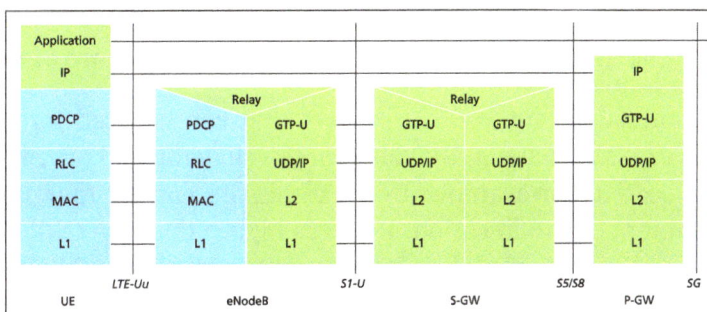

The E-UTRAN user plane protocol stack.

Data Handling during Handover

In the absence of any centralized controller node, data buffering during handover due to user mobility in the E-UTRAN must be performed in the eNodeB itself. Data protection during handover is a responsibility of the PDCP layer. The RLC and MAC layers both start afresh in a new cell after handover.

Control Plane

The protocol stack for the control plane between the UE and MME is shown in Figure. The blue region of the stack indicates the AS protocols. The lower layers perform the same functions as for the user plane with the exception that there is no header compression function for the control plane.

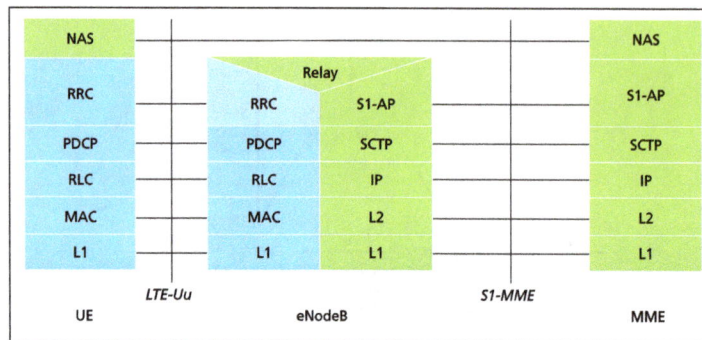

Control plane protocol stack.

The Radio Resource Control (RRC) protocol is known as "layer 3" in the AS protocol stack. It is the main controlling function in the AS, being responsible for establishing the radio bearers and configuring all the lower layers using RRC signaling between the eNodeB and the UE.

Quality of Service and EPS Bearers

In a typical case, multiple applications may be running in a UE at any time, each one having different quality of service requirements. For example, a UE can be engaged in a VoIP call while at the same time browsing a web page or downloading an FTP file. VoIP has more stringent requirements for QoS in terms of delay and delay jitter than web browsing and FTP, while the latter requires a much lower packet loss rate. In order to support multiple QoS requirements, different bearers are set up within the Evolved Packet System, each being associated with a QoS.

Broadly, bearers can be classified into two categories based on the nature of the QoS they provide:

- Minimum guaranteed bit rate (GBR) bearers that can be used for applications such as VoIP. These have an associated GBR value for which dedicated transmission resources are permanently allocated (for example, by an admission control function in the eNodeB) at bearer establishment or modification. Bit rates higher than the GBR may be allowed for a GBR bearer if resources are available. In such cases, a maximum bit rate (MBR) parameter, which can also be associated with a GBR bearer, sets an upper limit on the bit rate that can be expected from a GBR bearer.

- Non-GBR bearers that do not guarantee any particular bit rate. These can be used for applications such as web browsing or FTP transfer. For these bearers, no bandwidth resources are allocated permanently to the bearer.

In the access network, it is the responsibility of the eNodeB to ensure the necessary QoS for a bearer over the radio interface. Each bearer has an associated QCI, and an Allocation and Retention Priority (ARP).

Each QCI is characterized by priority, packet delay budget and acceptable packet loss rate. The QCI label for a bearer determines how it is handled in the eNodeB. Only a dozen such QCIs have been standardized so that vendors can all have the same understanding of the underlying service characteristics and thus provide corresponding treatment, including queue management, conditioning and policing strategy.

This ensures that an LTE operator can expect uniform traffic-handling behavior throughout the network regardless of the manufacturers of the eNodeB equipment. The set of standardized QCIs and their characteristics (from which the PCRF in an EPS can select) is provided in Table (from Section). The QCI table specifies values for the priority handling, acceptable delay budget and packet loss rate for each QCI label.

Table: Standardized QCIs for LTE.

QCI	Resource type	Priority	Packet delay budget (MS)	Packet error loss rate	Example services
1	GBR	2	100	10^{-2}	Conversational voice.
2	GBR	4	150	10^{-3}	Conversational video (live streaming).
3	GBR	5	300	10^{-6}	Non-conversational video (buffered streaming).
4	GBR	3	50	10^{-3}	Real-time gaming.
5	Non-GBR	1	100	10^{-6}	IMS signaling.
6	Non-GBR	7	100	10^{-3}	Voice, video (live streaming), interactive gaming.
7	Non-GBR	6	300	10^{-6}	Video (buffered streaming).
8	Non-GBR	8	300	10^{-6}	TCP-based (for example, WWW, e-mail), chat, FTP, p2p file sharing, progressive video and others.
9	Non-GBR	9	300	10^{-6}	

The priority and packet delay budget (and to some extent the acceptable packet loss rate) from the QCI label determine the RLC mode configuration and how the scheduler in the MAC handles packets sent over the bearer (for example, in terms of scheduling policy, queue management policy and rate-shaping policy). For example, a packet with higher priority can be expected to be scheduled before a packet with lower priority. For bearers with a low acceptable loss rate, an acknowledged mode (AM) can be used within the RLC protocol layer to ensure that packets are delivered successfully across the radio interface.

The ARP of a bearer is used for call admission control — that is, to decide whether or not the requested bearer should be established in case of radio congestion. It also governs the prioritization of the bearer for pre-emption with respect to a new bearer establishment request. Once successfully established, a bearer's ARP does not have any impact on the bearer-level packet forwarding treatment (for example, for scheduling and rate control). Such packet forwarding treatment should be solely determined by the other bearer level QoS parameters such as QCI, GBR and MBR.

An EPS bearer has to cross multiple interfaces as shown in Figure — the S5/S8 interface from the P-GW to the S-GW, the S1 interface from the S-GW to the eNodeB, and the radio interface (also known as the "LTE-Uu interface") from the eNodeB to the UE. Across each interface, the EPS bearer is mapped onto a lower layer bearer, each with its own bearer identity. Each node must keep track of the binding between the bearer IDs across its different interfaces.

An S5/S8 bearer transports the packets of an EPS bearer between a P-GW and an S-GW. The S-GW stores a one-to-one mapping between an S1 bearer and an S5/S8 bearer. The bearer is identified by the GTP tunnel ID across both interfaces.

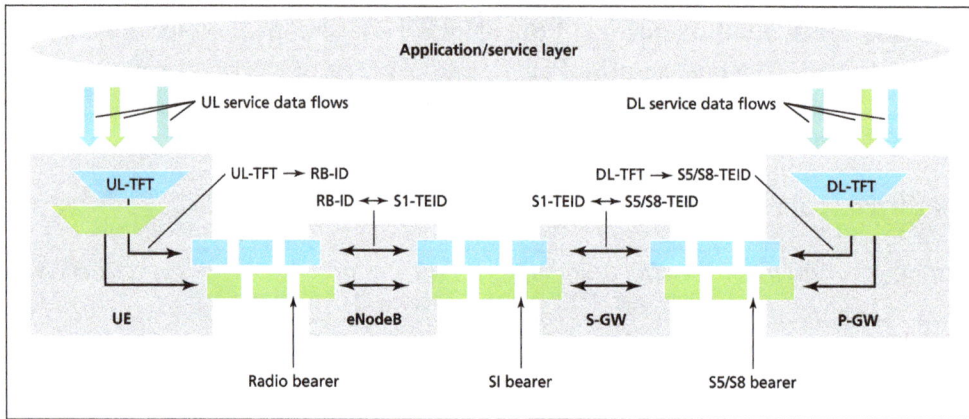

LTE/SAE bearer across the different interfaces.

The packets of an EPS bearer are transported by an S1 bearer between an S-GW and an eNodeB, and by a radio bearer between a UE and an eNodeB. An eNodeB stores a one-to-one mapping between a radio bearer ID and an S1 bearer to create the mapping between the two.

IP packets mapped to the same EPS bearer receive the same bearer-level packet forwarding treatment (for example, scheduling policy, queue management policy, rate shaping policy, RLC configuration). In order to provide different bearer-level QoS, a separate EPS bearer must therefore be established for each QoS flow. User IP packets must then be filtered into the appropriate EPS bearers.

Packet filtering into different bearers is based on TFTs. The TFTs use IP header information such as source and destination IP addresses and Transmission Control Protocol (TCP) port numbers to filter packets such as VoIP from web-browsing traffic, so that each can be sent down the respective bearers with appropriate QoS. An Uplink TFT (UL TFT) associated with each bearer in the UE filters IP packets to EPS bearers in the uplink direction. A Downlink TFT (DL TFT) in the P-GW is a similar set of downlink packet filters.

As part of the procedure by which a UE attaches to the network, the UE is assigned an IP address by the P-GW and at least one bearer is established. This is called the default bearer, and it remains established throughout the lifetime of the PDN connection in order to provide the UE with always-on IP connectivity to that PDN. The initial bearer-level QoS parameter values of the default bearer are assigned by the MME, based on subscription data retrieved from the HSS. The PCEF may change these values in interaction with the PCRF or according to local configuration. Additional bearers called dedicated bearers can also be established at any time during or after completion of the attach procedure. A dedicated bearer can be either a GBR or a non-GBR bearer (the default bearer always has to be a non-GBR bearer since it is permanently established). The distinction between default and dedicated bearers should be transparent to the access network (for example, E-UTRAN). Each bearer has an associated QoS, and if more than one bearer is established for a given UE, then each bearer must also be associated with appropriate TFTs. These dedicated bearers could be established by the network, based for example on a trigger from the IMS domain, or they could be requested by the UE. The dedicated bearers for a UE may be provided by one or more P-GWs.

The bearer-level QoS parameter values for dedicated bearers are received by the P-GW from the PCRF and forwarded to the S-GW. The MME only transparently forwards those values received from the S-GW over the S11 reference point to the E-UTRAN.

Bearer Establishment Procedure

This topic describes a typical end-to-end bearer establishment procedure across the network nodes, as shown in figure. When a bearer is established, the bearers across each of the interfaces discussed above are established.

The PCRF sends a Policy Control and Charging (PCC) Decision Provision message indicating the required QoS for the bearer to the P-GW. The P-GW uses this QoS policy to assign the bearer-level QoS parameters. The P-GW then sends a Create Dedicated Bearer Request message including the QoS and UL TFT to be used in the UE to the S-GW. After the S-GW receives the Create Dedicated Bearer Request message, including bearer QoS, UL TFT and S1-bearer ID, it forwards it to the MME.

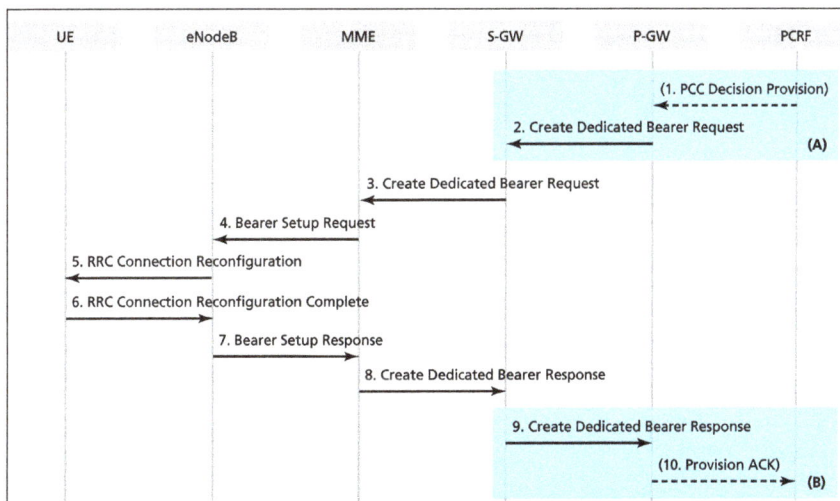

An example message flow for a LTE/SAE bearer establishment.

The MME then builds a set of session management configuration information including the UL TFT and the EPS bearer identity and includes it in the Bearer Setup Request message that it sends to the eNodeB. Since the session management configuration is NAS information, it is sent transparently by the eNodeB to the UE.

The Bearer Setup Request also provides the QoS of the bearer to the eNodeB; this information is used by the eNodeB for call admission control and also to ensure the necessary QoS by appropriate scheduling of the user's IP packets. The eNodeB maps the EPS bearer QoS to the radio bearer QoS and then signals an RRC Connection Reconfiguration message (including the radio bearer QoS, session management request and EPS radio bearer identity) to the UE to setup the radio bearer. The RRC Connection Reconfiguration message contains all the configuration parameters for the radio interface. These are mainly for the configuration of the layer 2 (the PDCP, RLC and MAC parameters), but also contain the layer 1 parameters required for the UE to initialize the protocol stack.

Messages 6 to 10 are the corresponding response messages to confirm that the bearers have been correctly set up.

The E-UTRAN Network Interfaces: S1 Interface

The provision of self-optimizing networks (SONs) is one of the key objectives of LTE. Indeed, self-optimization of the network is a high priority for network operators as a tool to derive the best performance from the network in a cost-effective manner, especially in changing radio propagation environments. Hence SON has been placed as a cornerstone of the LTE system from the beginning and is the concept around which all S1 and X2 procedures have been designed.

The S1 interface connects the eNodeB to the EPC. It is split into two interfaces, one for the control plane and the other for the user plane. The protocol structure for the S1 and the functionality provided over S1 are discussed in more detail below.

Protocol Structure Over S1

The protocol structure over S1 is based on a full IP transport stack with no dependency on legacy SS7 network configuration as used in GSM or UMTS networks. This simplification provides one expected area of savings on operational expenditure when LTE networks are deployed.

Control Plane

Figure shows the protocol structure of the S1 control plane, which is based on the well-known Stream Control Transmission Protocol / IP (SCTP/IP) stack.

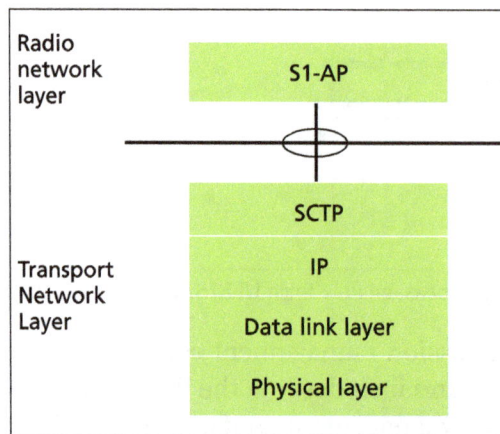

S1 control plane protocol stack.

The SCTP protocol is well known for its advanced features inherited from TCP that ensure the required reliable delivery of the signaling messages. In addition it makes it possible to benefit from improved features such as the handling of multi-streams to implement transport network redundancy easily and avoid head-of-line blocking or multi-homing.

An area of simplification in LTE (as compared to the UMTS Iu interface, for example) is the direct mapping of S1 Application Protocol (S1-AP) on top of SCTP. This results in a simplified protocol stack with no intermediate connection management protocol, since the individual connections are handled directly at the application layer. Multiplexing takes place between S1-AP and SCTP whereby each stream of an SCTP association is multiplexed with the signaling traffic of multiple individual connections.

LTE has also built flexibility into the lower layer protocols, giving the operator full optionality regarding the choice of IP version and the data link layer. For example, this enables the operator to start deployment using IP version 4 with the data link tailored to the network deployment scenario.

User Plane

Figure gives the protocol structure of the S1 user plane, based on the GTP/UDP5/IP stack, which is already well known from UMTS networks.

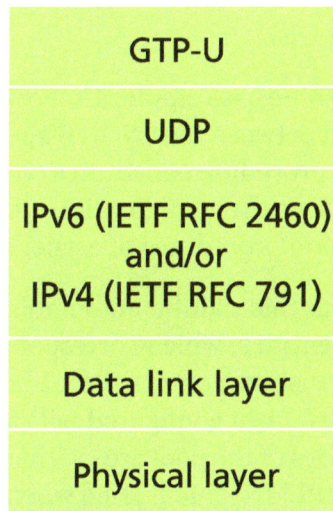

| GTP-U |
| UDP |
| IPv6 (IETF RFC 2460) and/or IPv4 (IETF RFC 791) |
| Data link layer |
| Physical layer |

S1-U user plane protocol stack.

One of the advantages of using GPRS Tunneling Protocol User plane (GTP-U) is its inherent facility to identify tunnels and to facilitate intra-3GPP mobility.

The IP version number and the data link layer have been left fully optional, as for the control plane stack.

A transport bearer is identified by the GTP tunnel endpoints and the IP address, destination TEID, source IP address, destination IP address). The S-GW sends downlink packets of a given bearer to the eNodeB IP address (received in S1-AP) associated to that particular bearer. Similarly, the eNodeB sends upstream packets of a given bearer to the EPC IP address (received in S1-AP) associated with that particular bearer.

Vendor-specific traffic categories (for example, real-time traffic) can be mapped onto Differentiated Services (DiffServ) code points (for example, expedited forwarding) by network operations and maintenance (O&M) configuration to manage QoS differentiation between the bearers.

Initiation Over S1

The initialization of the S1-MME control plane interface starts with the identification of the MMEs to which the eNodeB must connect, followed by the setting up of the Transport Network Layer (TNL).

With the support of the S1-flex function in LTE, an eNodeB must initiate an S1 interface toward each MME node of the pool area to which it belongs. The list of MME nodes of the pool together with an initial corresponding remote IP address can be directly configured in the eNodeB at deployment (although other means may also be used). The eNodeB then initiates the TNL establishment with that IP address. Only one SCTP association is established between one eNodeB and one MME.

During the establishment of the SCTP association, the two nodes negotiate the maximum number of streams that will be used over that association. However, multiple pairs of streams are typically used in order to avoid the head-of-line blocking issue. Among these pairs of streams, one must be reserved by the two nodes for the signaling of the common procedures (that is, those that are not specific to one UE). The other streams are used for the sole purpose of the dedicated procedures (that is, those that are specific to one UE).

Once the TNL has been established, some basic application-level configuration data for the system operation is automatically exchanged between the eNodeB and the MME through an S1 Setup procedure initiated by the eNodeB. This procedure constitutes one example of a network self-configuration process provided in LTE to reduce the configuration effort for network operators compared to the more usual manual configuration procedures of earlier systems.

An example of basic application data that can be automatically configured through the S1 Setup procedure is the tracking area identities, which correspond to the zones in which UEs will be paged. Their mapping to eNodeBs must remain consistent between the E-UTRAN and the EPC. Once the tracking area identities have been configured within each eNodeB, they are sent automatically to all the relevant MME nodes of the pool area within the S1 Setup Request message. The same applies for the broadcast list of PLMNs that is used when a network is being shared by several operators (each having its own PLMN-ID that needs to be broadcast for the UEs to recognize it). This saves a significant amount of configuration effort in the CN, avoids the risk of human error, and ensures that the E-UTRAN and EPC configurations regarding tracking areas and PLMNs are aligned.

Once the S1 Setup procedure has been completed, the S1 interface is operational.

Context Management Over S1

Within each pool area, a UE is associated with one particular MME for all its communications during its stay in this pool area. This creates a context in this MME for the UE. This particular MME is selected by the NAS Node Selection Function (NNSF) in the first eNodeB from which the UE entered the pool.

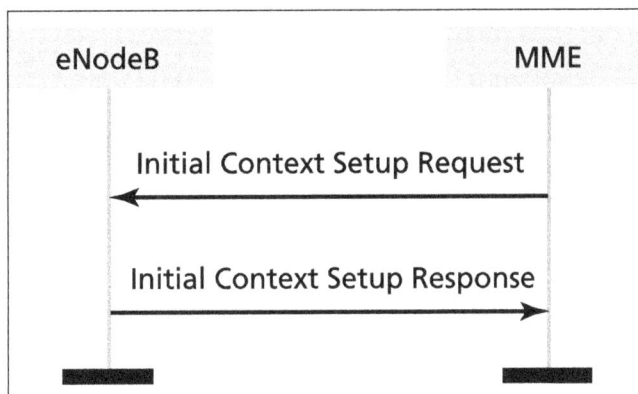

Initial Context Setup procedure.

Whenever the UE becomes active (that is, makes a transition from idle to active mode) under the coverage of a particular eNodeB in the pool area, the MME provides the UE context information to this eNodeB using the Initial Context Setup Request message. This enables the eNodeB in turn to create a context and manage the UE for the duration of its activity in active mode.

Even though the setup of bearers is otherwise relevant to a dedicated bearer management procedure described below, the creation of the eNodeB context by the Initial Context Setup procedure also includes the creation of one or several bearers including the default bearer.

At the next transition back to idle mode following a UE Context Release Command message sent from the MME, the UE context in the eNodeB is erased and only the UE context in the MME remains.

Bearer Management Over S1

LTE uses dedicated procedures independently covering the setup, modification and release of bearers. For each bearer requested to be setup, the transport layer address and the tunnel endpoint are provided to the eNodeB in the Bearer Setup Request message to indicate the termination of the bearer in the S-GW where uplink user plane data must be sent. Conversely, the eNodeB indicates in the Bearer Setup Response message the termination of the bearer in the eNodeB where the downlink user plane data must be sent. For each bearer, the QoS parameters requested for the bearer are also indicated. Independently of the standardized QCI values, it is also still possible to use extra proprietary QCI values for the fast introduction of new services if vendors and operators agree upon them.

Paging Over S1

In order to reestablish a connection toward a UE in idle mode, the MME distributes a paging request to the relevant eNodeBs based on the tracking areas where the UE is expected to be located. When it receives the Paging Request message, the eNodeB sends a page over the radio interface in the cells that are contained within one of the tracking areas provided in that message.

The UE is normally paged using its S-TMSI. The Paging Request message also contains a UE identity index value in order for the eNodeB to calculate the paging occasions at which the UE will switch on its receiver to listen for paging messages.

Mobility Over S1

LTE/SAE supports mobility within LTE/SAE as well as mobility to other systems using both 3GPP and non-3GPP technologies. The mobility procedures over the radio interface are described in topic. These mobility procedures also involve the network interfaces. The sections below discuss the procedures over S1 to support mobility.

Intra-LTE Mobility

There are two types of handover procedure in LTE for UEs in active mode: the S1 and X2 handover procedures.

For intra-LTE mobility, that is, mobility within the LTE system, the X2-handover procedure is normally used for inter-eNodeB handover. However, when there is no X2 interface between the two eNodeBs, or if the source eNodeB has been configured to initiate handover towards a particular target eNodeB through the S1 interface, then an S1-handover will be triggered.

The S1-handover procedure, shown in Figure, has been designed in a similar way to the UMTS Serving Radio Network Subsystem (SRNS) relocation procedure: it consists of a preparation phase, where the CN resources are prepared at the target side (steps 2 to 8), followed by an execution phase (steps 8 to 12) and a completion phase (after step 13).

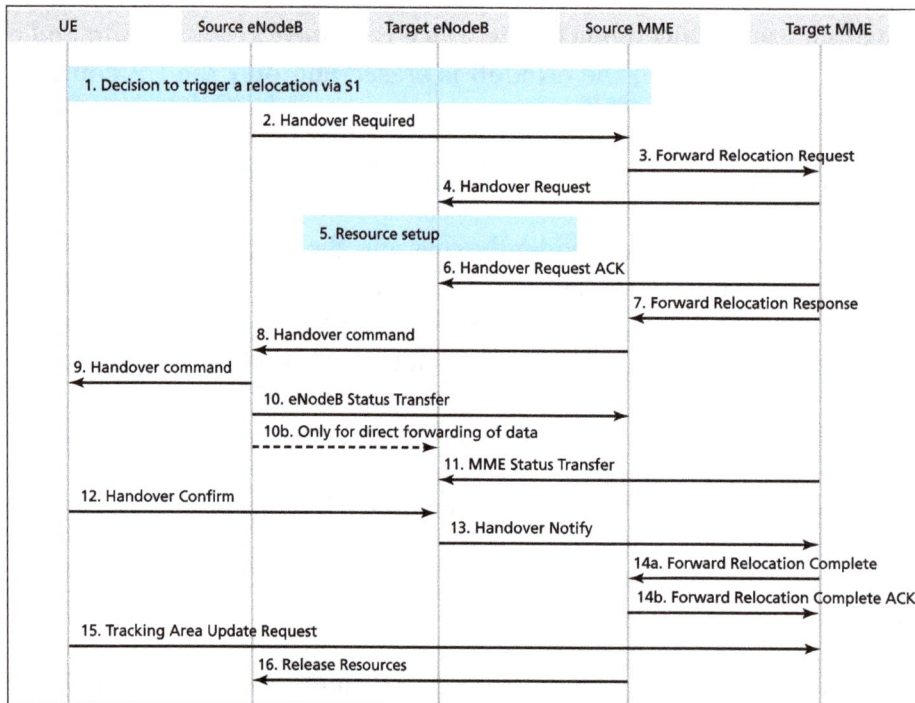

S1-based handover procedure.

Compared to UMTS, the main difference is the introduction of the Status Transfer message sent by the source eNodeB (steps 10 and 11). This message has been added in order to carry some PDCP status information that is needed at the target eNodeB in cases when PDCP status preservation applies for the S1-handover; this is in alignment with the information that is sent within the X2 Status Transfer message used for the X2-handover. As a result of this alignment, the handling of the handover by the target eNodeB as seen from the UE is exactly the same, regardless of the type of handover (S1 or X2) being used.

The Status Transfer procedure is assumed to be triggered in parallel with the start of data forwarding after the source eNodeB has received the handover command message from the source MME. This data forwarding can be either direct or indirect, depending on the availability of a direct path for the user plane data between the source eNodeB and the target eNodeB.

The Handover Notify message (step 13), which is sent later by the target eNodeB when the arrival of the UE at the target side is confirmed, is forwarded by the MME to trigger the update of the path switch in the S-GW toward the target eNodeB. In contrast to the X2-handover, the message is not acknowledged and the resources at the source side are released later upon reception of a Release Resource message directly triggered from the source MME.

Inter-radio Access Technology Mobility

One key element of the design of the first release of LTE is the need to co-exist with other technologies.

For mobility from LTE toward UMTS, the handover process can reuse the S1-handover procedures described above, with the exception of the Status Transfer message, which is not needed at steps 10 and 11 since no PDCP context is continued.

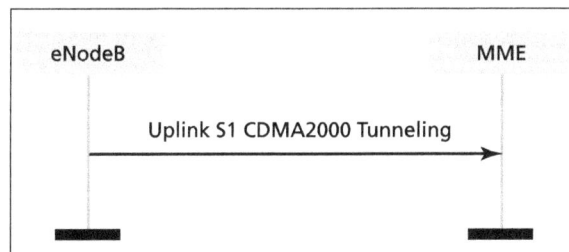

Uplink S1 CDMA2000 tunneling procedure.

For mobility toward CDMA2000, dedicated uplink and downlink procedures have been introduced in LTE. They essentially aim at tunneling the CDMA2000 signaling between the UE and the CDMA2000 system over the S1 interface, without being interpreted by the eNodeB on the way. The Uplink S1 CDMA2000 Tunneling message shown in Figure also includes the Radio Access Technology (RAT) type in order to identify the CDMA2000 RAT with which the tunneled CDMA2000 message is associated in order for the message to be routed to the correct node within the CDMA2000 system.

Load Management Over S1

Three types of load management procedures apply over S1: a normal load balancing procedure to distribute the traffic, an overload procedure to overcome a sudden peak in the loading and a load rebalancing procedure to partially/fully offload an MME.

The MME load balancing procedure aims to distribute the traffic to the MMEs in the pool evenly according to their respective capacities. To achieve this goal, the procedure relies on the normal NNSF present in each eNodeB as part of the S1-flex function. Provided that suitable weight factors corresponding to the capacity of each MME node are available in the eNodeBs beforehand, a weighted NNSF done by each and every eNodeB in the network normally achieves a statistically balanced distribution of load among the MME nodes without further action. However, specific actions are still required for some particular scenarios:

- If a new MME node is introduced (or removed), it may be necessary temporarily to increase (or decrease) the weight factor normally corresponding to the capacity of this node in order to make it catch more (or less) traffic at the beginning until it reaches an adequate level of load.

- In case of an unexpected peak in the loading, an Overload message can be sent over the S1 interface by the overloaded MME. When received by an eNodeB, this message calls for a temporary restriction of a certain type of traffic. An MME can adjust the reduction of traffic it desires by defining the number of eNodeBs to which it sends the Overload message and by defining the types of traffic subject to restriction.

- Finally, if the MME wants to rapidly force the offload of some or all of its UEs, it will use the rebalancing function. This function forces the UEs to reattach to another MME by using a specific "cause value" in the UE Release Command S1 message. In a first step it applies to idle mode UEs and in a second step it may also apply to UEs in connected mode (if full MME offload is desired, for example, for maintenance reasons).

The E-UTRAN Network Interfaces: X2 Interface

The X2 interface is used to interconnect eNodeBs. The protocol structure for the X2 interface and the functionality provided over X2 are discussed below.

Protocol Structure Over X2

The control and user plane protocol stacks over the X2 interface, shown in figures respectively, are the same as those for the S1 interface, with the exception that X2-AP is substituted for S1-AP. This also reaffirms that the choice of the IP version and the data link layer are fully optional.

The use of the same protocol structure over both interfaces provides advantages such as simplifying the data forwarding operation.

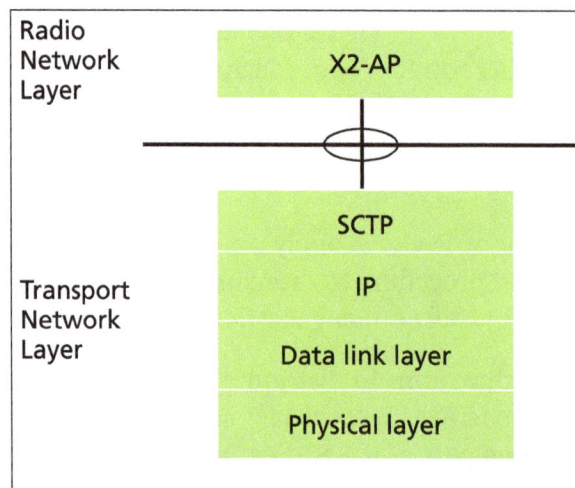

X2 signaling bearer protocol stack.

GTP-U
UDP
IPv6 (IETF RFC 2460) and/or IPv4 (IETF RFC 791)
Data link layer
Physical layer

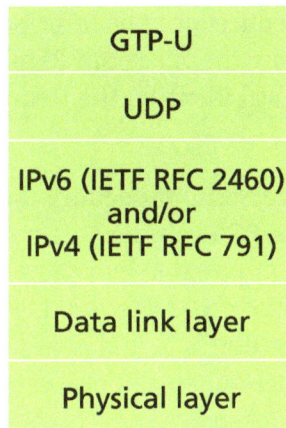

Transport Network Layer for data streams over X2.

Initiation Over X2

The X2 interface may be established between one eNodeB and some of its neighbor eNodeBs in order to exchange signaling information when needed. However, a full mesh is not mandated in an E-UTRAN network. Two types of information may typically need to be exchanged over X2 to drive the establishment of an X2 interface between two eNodeBs: load- or interference-related information and handover-related information.

Because these two types of information are fully independent of one another, it is possible that an X2 interface may be present between two eNodeBs for the purpose of exchanging load or interference information, even though the X2-handover procedure is not used to handover UEs between those eNodeBs. In such a case, the S1-handover procedure is used instead.

The initialization of the X2 interface starts with the identification of a suitable neighbor followed by the setting up of the TNL.

The identification of a suitable neighbor may be done by configuration or alternatively by a function known as the automatic neighbor relation function (ANRF). This function makes use of the UEs to identify the useful neighbor eNodeBs: an eNodeB may ask a UE to read the global cell identity from the broadcast information of another eNodeB for which the UE has identified the physical cell identity (PCI) during the new cell identification procedure. The ANRF is another example of a SON process introduced successfully in LTE. Through this self-optimizing process, UEs and eNodeB measurements are used to auto-tune the network.

Once a suitable neighbor has been identified, the initiating eNodeB can set up the TNL using the X2 IP address of this neighbor, either as retrieved from the network or locally configured. In particular, a SON-configuration dedicated procedure over S1 termed the eNB Configuration Transfer procedure has been designed to enable the initiating eNodeB to directly request over the S1 interface the X2 IP address of a discovered neighbor eNodeB to be used for X2 establishment. This network solution through the S1 interface may avoid the need for an operator to use other more complex network solutions such as the deployment of DNS servers.

After the TNL has been set up, the initiating eNodeB must trigger the X2 Setup procedure. This procedure enables an automatic exchange of application level configuration data relevant to the X2 interface, similar to the S1 Setup procedure already described in Section For example, each eNodeB reports to a neighbor eNodeB, using the X2 Setup Request message, information about each cell it manages, such as the cell's physical identity, the frequency band, the tracking area identity and/or the associated PLMNs.

This automatic data exchange in the X2 Setup procedure is also the core of another SON feature: the automatic self-configuration of the PCIs. Under this new SON feature, the O&M system can provide the eNodeBs with either a list of possible PCI values to use or a specific PCI value. In the first case, in order to avoid collisions, the eNodeB should use a PCI that is not already used in its neighborhood. Because the PCI information is included in the LTE X2 Setup procedure, while detecting a neighbor cell by the ANR function, an eNodeB can also discover all the PCI values used in the neighborhood of that cell and consequently eliminate those values from the list of suitable PCIs to start with.

Once the X2 Setup procedure has been completed, the X2 interface is operational.

Mobility Over X2

Handover through the X2 interface is triggered by default for intra-LTE mobility unless there is no X2 interface established or the source eNodeB is configured to use S1-handover instead. The X2-handover procedure is illustrated in Figure. Like S1-handover, it is also composed of a preparation phase (steps 4 to 6), an execution phase (steps 7 to 9) and a completion phase (after step 9).

The key features of X2-handover for intra-LTE handover are:

- The handover is directly performed between two eNodeBs, making the preparation phase quick.

- Data forwarding may be operated per bearer in order to minimize data loss.

- The MME is only informed at the end of the handover procedure when the handover is successful, in order to trigger the path switch.

- The release of resources at the source side is directly triggered from the target eNodeB.

For those bearers requiring in-sequence delivery of packets, the Status Transfer message (step 8) provides the sequence number (SN) and the Hyper Frame Number (HFN) that the target eNodeB should assign to the first packet with no SN yet assigned that it must deliver. This first packet can either be one received over the target S1 path or one received over X2, if data forwarding over X2 is used. When the source eNodeB sends the Status Transfer message, it freezes its transmitter/receiver status, that is, it stops assigning PDCP SNs to downlink packets and stops delivering uplink packets to the EPC.

Mobility over X2 can be categorized according to its resilience to packet loss: the handover can be termed "seamless" if it minimizes the interruption time during the move of the UE or "lossless" if it tolerates no loss of packets at all. These two modes use data forwarding of user plane downlink packets. The source eNodeB may decide to operate one of these two modes on a per-EPS-bearer basis, based on the QoS received over S1 for this bearer and the service at stake.

Seamless Handovers

If the source eNodeB selects the seamless mode for one bearer, it proposes to the target eNodeB in the Handover Request message to establish a GTP tunnel to operate the downlink data forwarding. If the target eNodeB accepts, it indicates in the Handover Request ACK message the tunnel endpoint where the forwarded data is expected to be received. The tunnel endpoint may be different from the one set up as the termination point of the new bearer established over the target S1.

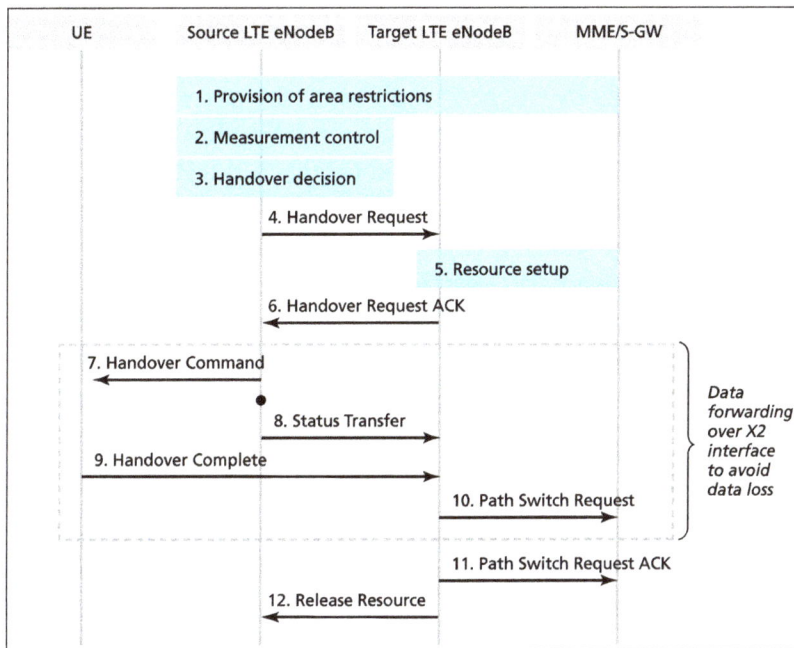

X2-based handover procedure.

Upon receipt of the Handover Request ACK message, the source eNodeB can start forwarding the freshly arriving data over the source S1 path toward the indicated tunnel endpoint in parallel to sending the handover trigger to the UE over the radio interface. The forwarded data is thus available at the target eNodeB to be delivered to the UE as early as possible.

When forwarding is in operation and in-sequence delivery of packets is required, the target eNodeB is assumed to first deliver the packets forwarded over X2 before delivering the ones received over the target S1 path, once the S1 path switch has been done. The end of the forwarding is signaled over X2 to the target eNodeB by the reception of "special GTP packets" that the S-GW has inserted over the source S1 path just before switching this S1 path; these are then forwarded by the source eNodeB over X2 like any other regular packets.

Lossless Handovers

If the source eNodeB selects the lossless mode for one bearer, it will additionally forward over X2 those user plane downlink packets that it has PDCP-processed but that are still buffered locally because they have not yet been delivered and acknowledged by the UE. These packets are forwarded together with their assigned PDCP SN included in a GTP extension header field. They are sent over X2 prior to the fresh arriving packets from the source S1 path. The same mechanisms described above for the seamless handover are used for the GTP tunnel establishment. The end of forwarding is also handled in the same way, since in-sequence packet delivery applies to lossless handovers. In addition, the target eNodeB must ensure that all the packets, including the ones received with sequence number over X2, are delivered in sequence to the target side.

A new feature in LTE is the optimization of the radio by selective retransmission. When lossless handover is used, the target eNodeB may not deliver over the radio interface some of the forwarded downlink packets received over X2 if it is informed by the UE that these packets have already been received at the source side. This is called downlink selective retransmission.

Similarly in the uplink, the target eNodeB may not wish the UE to retransmit packets already received earlier at the source side by the source eNodeB, for example, to avoid wasting radio resources. To use uplink selective retransmission, the source eNodeB forwards the user plane uplink packets that it has received out of sequence to the target eNodeB, over a new GTP tunnel. The target eNodeB must first request the source eNodeB to establish the new forwarding tunnel by including a GTP tunnel endpoint where it expects the forwarded uplink packets to be received in the Handover Request ACK message. If possible, the source eNodeB then indicates in the Status Transfer message for this bearer, the list of SNs corresponding to the expected forwarded packets. This list helps the target eNodeB inform the UE earlier of the packets that are not to be retransmitted, making the overall uplink selective retransmission scheme faster.

Multiple Preparations

Another new feature of the LTE handover procedure compared to UMTS is multiple preparation. This feature enables the source eNodeB to trigger the handover preparation procedure towards multiple candidate target eNodeBs. Even though only one of the candidates is indicated as target to the UE, this makes recovery faster in case the UE fails to attach to the target and connects to one of the other prepared candidate eNodeBs instead. The source eNodeB receives only one Release Resource message from the final selected eNodeB.

Regardless of whether multiple or single preparation is used, the handover can be canceled during or after the preparation phase. If the multiple preparation features is used, it is for example recommended that upon reception of the Release Resource message the source eNodeB triggers a Cancel procedure toward each of the non-selected prepared eNodeBs.

Load and Interference Management Over X2

The exchange of load information between eNodeBs is of key importance in the flat architecture used in LTE, as there is no central RRM node as was the case, for example, in UMTS with the Radio Network Controller (RNC). The exchange of load information falls into two categories depending on the purpose it serves.

- The exchange of load information can be used for the X2 load balancing process, in which case the relevant frequency of exchange is rather low (in the order of seconds);

- The exchange of load information can be used to optimize some RRM processes, such as interference coordination, in which case the frequency of exchange is rather high (in the order of tens of milliseconds).

Load Balancing

Like the ANRF SON function described in Section, load balancing is another aspect of SON built into the design of LTE. The objective of load balancing is to counteract local traffic load imbalance between neighboring cells with the aim of improving the overall system capacity. One solution is to optimize the cell reselection/handover parameters (such as thresholds and hysteresis) between neighboring cells autonomously upon detection of an imbalance.

In order to detect an imbalance, itis necessary to compare the load of the cells and therefore to exchange information about them between neighboring eNodeBs. The cell load information exchanged can be of different types: radio measurements corresponding to the usage of physical resource blocks, possibly partitioned into real-time and non-real-time traffic; or generic measurements representing non-radio-related resource usage such as processing or hardware occupancy.

A client-server mechanism is used for the load information exchange: the Resource Status Response and Update messages are used to report the load information over the X2 interface between a requesting eNodeB (client) and the eNodeBs that have subscribed to this request (servers). The reporting of the load is periodic and according to the periodicity expressed in the Resource Status Request message that triggered the procedure.

Interference Management

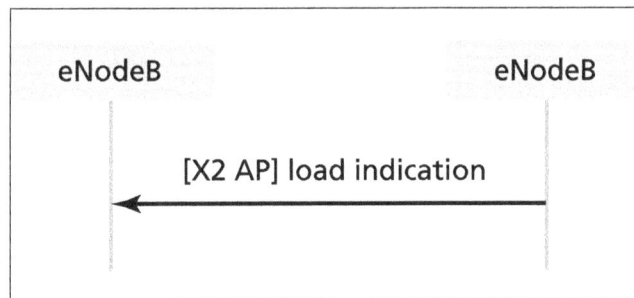

Load indication over X2 interface.

A separate load indication procedure is used over the X2 interface for the exchange of load information related to interference management as shown in Figure. As these measurements have direct influence on some RRM real-time processes, the frequency of reporting using this procedure may be high.

For the uplink interference, two indicators can be provided within the load indication message: a "High Interference Indicator" and an "Overload Indicator."

UE Historical Information Over X2

Historical UE information constitutes another example of a feature designed to support SON built into the design of LTE. It is embedded within the X2-handover procedure. Historical UE information consists, for example, of the last few cells visited by the UE, together with the time spent in each one. This information is propagated from one eNodeB to another within the Handover Request messages and can be used to determine the occurrence of ping-pong between two or three cells for instance. The length of the history information can be configured for more flexibility.

More generally, the historical UE information consists of some RRM information that is passed from the source eNodeB to the target eNodeB within the Handover Request message to assist the RRM management of a UE or of a cell. The information can be partitioned into two types:

- UE RRM-related information passed over X2 within the RRC transparent container.

- Cell RRM-related information, passed over X2 directly as an information element of the X2-AP Handover Request message itself.

LTE-Advanced Mobile Network Plan Layout

The core network of the LTE-Advanced system is separated into many parts. Figure shows how each component in the LTE-Advanced network is connected to one another. NodeB in 3G system was replaced by evolved NodeB (eNB), which is a combination of NodeB and radio network controller (RNC). The eNB communicates with User Equipments (UE's) and can serve one or several cells at one time. Home eNB (HeNB) is also considered to serve a femtocell that covers a small indoor area. The evolved packet core (EPC) comprises of the following four components. The serving gateway (SGW) is responsible for routing and forwarding packets between UE's and packet data network (PDN) and charging. In addition, it serves as a mobility anchor point for handover. The mobility management entity (MME) manages UE access and mobility, and establishes the bearer path for UE's. Packet data network gateway (PDN GW) is a gateway to the PDN, and policy and charging rules function (PCRF) manages policy and charging rules.

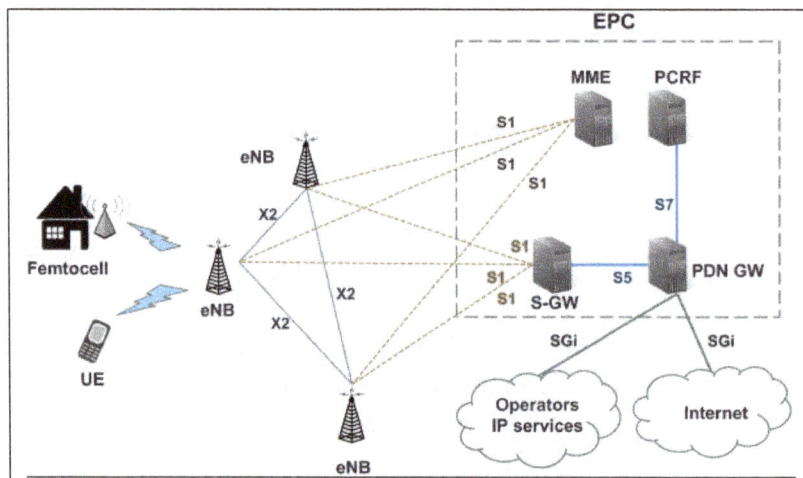

Network architecture of LTE-Advanced.

Mobile terminal location, can be outdoor or indoor. If the mobile terminal is located inside buildings the environment is called indoor, otherwise it is outdoor.

Antenna location, it can be above or below the average rooftop level. In case, when base station antenna array is above average height of the buildings, the environment is considered to be macro-cellular and, in case, when base station antenna array is below average height of the buildings, the environment is considered to be microcellular. There is even smaller type of the cells than macro and micro cells, so called pico cells for which the antennas are located mainly in indoor environments if it located in shopping mall or enterprise. In Femto cells, the antennas are located mainly in indoor environments if it located in home.

Morphography type, urban, suburban, rural. These area types are determined by the variation of size and density of both manmade and natural obstacles located in surroundings of User Equipment (UE) and base station sites.

Figure illustrates types of cell according to the surrounding environments.

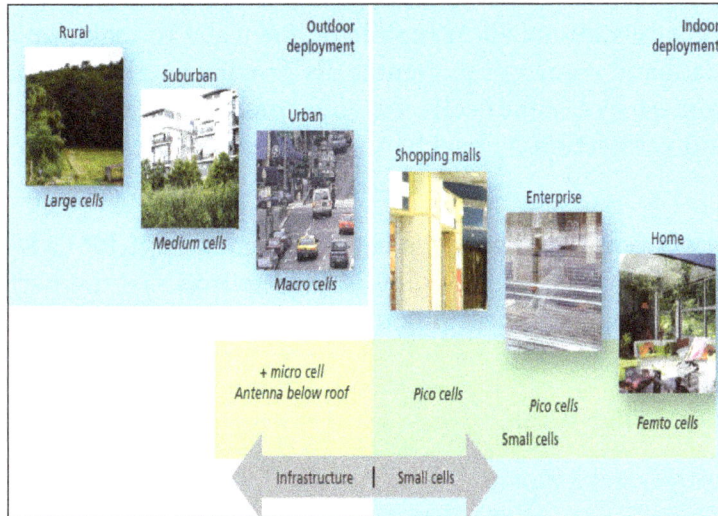

Radio propagation environments.

A Heterogeneous Network (HetNet) is a mix of highpower macro-eNBs responsible for umbrella coverage mainly for outdoor users, and low-power micro/Pico/Femto/relay BSs that are deployed for incremental capacity growth and coverage enhancement.

Macrocells: A Macrocell provides the largest area of coverage within a mobile network. Its antennas can be mounted on ground-based masts, rooftops or other structures and must be high enough to avoid obstruction. Macrocells provide radio coverage over varying distances, depending on the frequency used, the number of calls and the physical terrain. Typically they have a power output in tens of watt. Macrocells are conventional base stations with power about 20W, that use dedicated backhaul, are open to public access and range is about 1 km to 20 km.

Table: Comparison of Small Cells.

Cell type	Typical Cell Size	Data Rate Limitation
Macro	1-30 km	Propagation.
Micro	500 m-2 km	Capacity and propagation.
Pico	4-200 m	Capacity and propagation.
Femto	10 m	Broadband connection and Handset.

Microcells: Microcells provide additional coverage and capacity in areas where there are high numbers of users, for Example, urban and suburban areas. Microcells cover around10% of the area of a Macrocell. The antennas for microcells are mounted at street level, are smaller than Macrocell antennas and can often be disguised as building features so that they are less visually intrusive. Microcells have lower output powers than marocells, usually a few watts. Microcells are base stations with power between 1 to 5W, that use dedicated backhaul, are open to public access and range is about 500 m to 2 km.

Picocells: Picocells provide more localized coverage. These are generally found inside buildings where coverage is poor or where there is a dense population of users such as in airport terminals, train stations and shopping centers. Picocells are low power base stations with power ranges from 50 mW to 1W that use dedicated backhaul connections; open to public access and range is about 200 m or less.

Femtocells: Femtocell base stations allow mobile phone users to make calls inside their homes via their Internet broadband connection. Femtocells provide small area coverage solutions operating at low transmit powers. Femtocells are consumer deployable base stations that utilize consumer's broadband connection as backhaul, may have restricted association and power is less than 100 mW.

Relay Node (RN): For efficient heterogeneous network planning, 3GPP LTE-Advanced has introduced concept of Relay Nodes (RNs). Relaying is used to improve the performance of LTE, in terms of coverage and throughput. According to 3GPP, the use of relays will allow the following improvements:

- Provide coverage in new areas.

- Temporary network deployment.

- Cell-edge throughput.

- Coverage of high data rate.

- Group mobility.

- Cost reduction: The cost of a relay, by itself, should be less than the cost of an eNB, assuming that the complexity of a relay is less than the complexity of an eNB. Due to the lack of a wired backhaul, the deployment cost and time should also be reduced, compared to an eNB.

- Power consumption reduction: The single-hop distance between the eNB and the UE is divided into two distances: the distance from the eNB to the relay, and the distance from the relay to the UE. Cost reduction: The cost of a relay, by itself, should be less than the cost of an eNB, assuming that the complexity of a relay is less than the complexity of an eNB. Due to the lack of a wired backhaul, the deployment cost and time should also be reduced, compared to an eNB.

In figure, the basic scheme in which relays are planned to be deployed in LTE-Advanced is depicted.

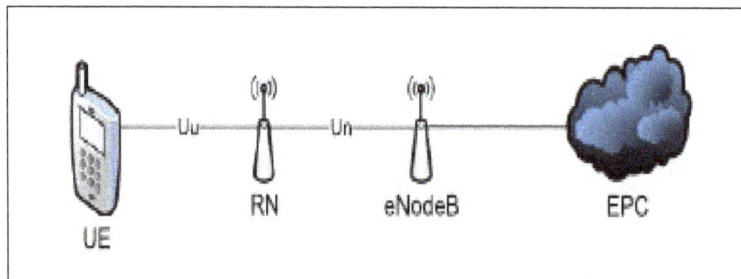

Relay basic scheme.

MIMO Techniques

Multi-antenna or MIMO (Multiple Input, Multiple Output) technology is based on transmitting and receiving with multiple antennas and utilizing uncorrelated communication channels when radio signals propagate through the physical environment. If there is enough isolation between the communication channels, then multiple data transmissions can share the same frequency resources. If the multiple transmissions are for a single user, then the technology is called Single-User MIMO (SU-MIMO), for multiple users Multi-User MIMO (MUMIMO).

The better the system can utilize these communication channels for multiple transmissions, the higher is the capacity that the system can provide.

MIMO performance is subject to a large number of parameters: the number of transmitter and receiver antennas, reference signals and algorithms for channel estimation, feedback of channel estimation data from the receiver to the transmitter and spatial encoding methods. Consequently a comprehensive design is crucial to provide optimum system performance.

LTE-A Network Planning

To be able to plan and implement a cost efficient high quality cellular mobile wireless network, very careful radio network planning procedure must be done. Thus, the planning process carried out in phases and each phase is well documented. The radio network planning procedure requires good knowledge about the coverage area, propagation environment, traffic load and required services to be able to analyses the network and to decide the optimal radio network planning strategy. The fact, that all the above mentioned aspects are not constant and vary in time, makes the radio network planning a nonstop process, which requires continuous monitoring and optimization.

The radio network planning is a process, which defines different steps, like measurements, planning, documentation, etc. that should be done in different phases to manage connections between coverage, capacity and interference. The coverage or capacity or QoS is not possible to maximize simultaneously, but all of them need to be optimized in order to implement a cost-efficient high quality radio network. To provide necessary coverage and, at the same time, optimize capacity and quality, the radio network planning can be divided into three main phases, illustrated in Figure. These phases can be used from initial deployment of the radio networks to their evolution and further development.

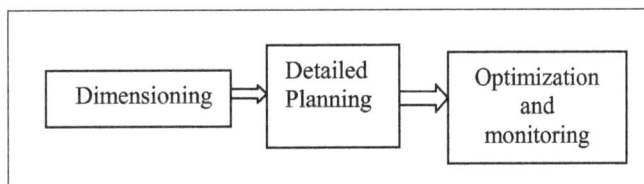

Radio System Planning Process Phases.

During the first phase, dimensioning, the planned network configuration is analyzed and an appropriate radio network deployment strategy is defined. In second phase, detailed planning, the detailed design and actual implementation of the radio network is done. First step in detailed planning is the configuration planning, which need to be done prior to coverage and capacity planning to be able to analyses all available coverage and capacity related software and hardware features.

The base station site configuration, which is different for different environments, need to be done based on both coverage and capacity requirements. Coverage specific requirements define coverage related base station elements and capacity requirements define capacity related base station elements. And finally, power budget can be calculated based on optimized base station parameters. Eventually, the configuration planning will provide total base station site configuration for different places and environments.

The configuration planning is followed by coverage planning, the aim of which is to minimize the number of base station sites by utilizing output information of the dimensioning and configuration planning. An important role in configuration planning plays surveying which helps to find out potential propagation problems and suggest base station sites locations. After that, some measurements can be done to tune propagation models for the particular areas. The tuned propagation models will give the final locations for base stations, by taking as input base station configuration parameters as well as some information about environment. The final coverage prediction and base station locations are usually defined by the use of advance planning software.

The next step is capacity planning which should be started as soon as the base station sides are selected. The capacity planning is done by the use of planning-tools, as the resource allocation mechanisms are already defined in dimensioning phase. The initial step is to define planning thresholds, after that, the main job will be done by planning-tools. The last step in the detailed planning is parameter planning, which is done immediately before the launch of the network.

The last step in the detailed planning is parameter planning, which is done immediately before the launch of the network. The last step in the detailed planning is parameter planning, which is done immediately before the launch of the network.

The last step in the detailed planning is parameter planning, which is done immediately before the lunch of the network. Radio performance has a direct impact on the cost of deploying the network in terms of the required number of base station sites and in terms of the transceivers required. The operator is interested in the network efficiency: how many customers can be served, how much data can be provided and how many base station sites are required. The efficiency is considered in the link budget calculations and in the capacity simulations. The end user application performance depends on the available bit rate, latency and seamless mobility. The radio performance defines what applications can be used and how these applications perform. Recent works in LTE and LTE-A network planning are divided into two directions. The first direction is solving Capacity and Coverage optimization by Self-Organizing LTE network. A novel hybrid two layer optimization framework is proposed to enhance the network capacity and coverage, where on the top layer a network entity of eCoordinator is implemented to ensure overall network coverage by optimizing the antenna tilt and capacity-coverage weight of each cell in a centralized manner, and on the bottom layer individual eNB optimizes cell-specific capacity and coverage by tuning its pilot power in a distributed manner. A heuristic algorithm is developed for the eCoordinator operation at large time granularity and the Genetic Programming (GP) approach is exploited for the eNB operation at small time granularity, for the purpose of tracking overall network performance as well as adapting to network dynamics. Results have demonstrated the usefulness of the proposed algorithms by enhancing network capacity and coverage performance under various system requirements.

Present reinforcement learning strategies for selforganized coverage and capacity optimization through antenna down tilt adaptation. This work analyzes different learning strategies for a Fuzzy Q-Learning based solution in order to have a fully autonomous optimization process. The learning behavior of these strategies is presented in terms of their learning speed and convergence to the optimal settings. Simultaneous actions by different cells of the network have a great impact on this learning behavior. Therefore, a study for stable strategy where only one cell can take an action per network snapshot as well as a more dynamic strategy where all the cells take simultaneous actions in every snapshot, also propose a cluster based strategy that tries to combine the benefits of both. The performance is evaluated in all three different network states, i.e. deployment, normal operation and cell outage. The simulation results show that the proposed cluster based strategy is much faster to learn the optimal configuration than one-cell-per-snapshot and can also perform better than the all-cells-per-snapshot strategy due to better convergence capabilities.

From the discussion of architecture of LTE-A, we can divide the network planning of LTE-A into main type as:

- Indoor network planning.

- Outdoor network planning.

Indoor Network Planning

Small cells offer mobile service providers (MSPs) a cost-effective alternative to macro-only deployments for meeting growing coverage and capacity demands. That's because as small, low-cost access points, they are selfinstalled (home and enterprise cells) or easily installed by a single person (metro cells). Plus, as small cells are added, they offload traffic from the macro network. This increases available network capacity without the deployment of new macro sites. Owned and managed by the MSP —metro cells, small cells — are most cost effective in areas where new macro sites are required. The larger the number of macro sites, the greater the economic benefits. Metro cells cost much less than macro radio equipment and they do not require civil works that contribute heavily to macro site deployment costs.

There are several factors behind the trend to smaller cells, including the perceived risks to health and visual appearance. Larger cells that transmit more radio waves sometimes spark concerns about radiation, while at the same time are held to be less aesthetically pleasing, especially in dense locations. Smaller cells also consume less power, reducing energy demands and offering potential environmental benefits. As the number of cells rises along with the demand for anywhere, anytime access, mobile service providers face a major challenge: How to define and deliver high-quality services cost-efficiently, and how to address the corresponding infrastructure and management challenges. Meeting this challenge means looking at the full set of requirements from a solutions lifecycle perspective, beginning with architecture and finishing with deployment. Designing for small-cell environments near and in-building can be a daunting task. There is a host of legal, logistical, technological, and other issues to consider from the outset. Nonetheless, quality of design is key to creating a sustainable In-Building solution spanning time, location and mobile generations.

In terms of architecture, LTE introduces new concerns and is more complex than 2G or 3G. Capacity requirements must be carefully considered, along with the impact of the macro network. Moreover, new antenna features such as MIMO and Beam Forming have to be taken into account, as well as end-to-end planning, integration and validation of IP networking and applications. Given the range and complexity of these issues, solution architects need to create an end-to-end reference architecture document detailing the necessary products and the interconnectivities among different elements and subsystems of the LTE network. The Solution Architect must also deliver a highlevel design and well-documented technical interfaces. Technical deliverables must be reviewed to assure consistency with solution architecture, customer requirements, and quality goals. All of this is undertaken in keeping with the mission of the Solution Architect to mitigate delivery risk through careful scrutiny of the scope of work and clear communications. Low-power base stations such as femtocells are one of the candidates for high-data-rate provisioning in local areas, such as residences, apartment complexes and business offices. Due to the expected large number of user-deployed cells, centralized network planning becomes impractical, and new scalable alternatives must be sought.

It relies on a suite of simulation tools including generation of random 3D femto-cell deployments in real environments, realistic path-loss predictions using a raybased model and a 3D downlink performance analysis (i.e. considering all floors) of heterogeneous LTE networks in terms of coverage, macro offload and throughput. A first study demonstrates a significant macro offload and power consumption reduction in a realistic dense corporate FAP deployment. Then, a second study shows the large growth of indoor capacity enabled by dense FAP deployments but also the coverage degradation for non-subscribers when closed-access mode is used.

Outdoor Network Planning

The second direction is the study of Energy and cost impacts of relay and femtocells deployments in long-termevolution advanced where presents a methodology for estimating the total energy consumption, taking into account the total operational power and embodied energy, and TCO of wireless cellular networks, and in particular provides a means to compare homogeneous and heterogeneous network (HetNets) deployments. The realistic energy models and energy metrics based on information available from mobile-network operators (MNOs) and base stations manufacturers must taking into consideration. Additionally, up-to-date operational and capital expenditure (OPEX and CAPEX) models are used to calculate TCO of candidate networks. There are two scenarios for HetNets, namely a joint macro-relay network and a joint macro–femtocell network, with different relay and femtocell deployments densities. The results obtained show that compared to macro-centric networks, joint macro-relay networks are both energy and cost efficient, whereas joint macro–femtocell networks reduce the networks TCO at the expense of increased energyconsumption. Finally, it is observed that energy and cost gains are highly sensitive to the OPEX model adopted.

References

- J. Ghosh and s. Dhar roy, "qualitative analysis for coverage probability and energy efficiency in cognitive-femtocell networks under macrocell infrastructure", electronics letters, vol. 51,no. 17,pp.1378–1380,aug.2015. Doi:10.1049/el.2015.0171

- What-is-a-cell-tower, mobile-technology: whatsag.com, Retrieved 5 February, 2019

- At&t 3g microcell - wireless signal booster". Wireless from at&t. Archived from the original on 2010-02-21. Retrieved 2012-07-26

- Macrocell, definition: techopedia.com, retrieved 31 march, 2019

- What-is-a-cell-phone-signal-booster-and-how-does-it-work, pages: signalbooster.com, retrieved 14 july, 2019

Mobile Network: Threats, Attacks and Security

5

- **Mobile Network Threats**
- **Signalling Attacks in Mobile Networks**
- **Attacks on LTE in 4G**
- **IP Spoofing in Mobile Network**
- **Mobile Device Security**
- **Security Threat and Counter-measure on 3G Network**
- **Security of Software Defined Mobile Networks**

Mobile network is prone to different kinds of threats with the increasing number of users and operators. Data leakage, network spoofing, phishing attacks, broken cryptography, etc. are some of the threats to network security. This chapter has been carefully written to provide an in-depth understanding of these threats, attacks and security of mobile networks.

Mobile Network Threats

Initial CDPD offerings by 1G operator and subsequent rollouts of GPRS, EDGE, CDMA, and 3G services such as CDMA EV-DO and UMTS/HSDPA, 1 create an all IP-based environment that makes it easier to attack the infrastructure. Attackers have readily available tools to assault these networks, and with the availability of USB modems for 3G access, they have a simple path to threaten any mobile network. Data flows are difficult to analyze and even valid connections can pose a threat – for example, when too many requests are sent legitimately.

There is an additional threat to external networks that are now allowed to connect with systems within the mobile network. Previously, these networks were closed to data connections from the outside. These new connections have to be verified – stateful inspection is required to prevent unauthorized packets from flowing into the network.

Different Kinds of Attacks

Threats to mobile networks include, but are not limited to, information confidentiality, data integrity, and service availability. Below are various types of attacks:

Traditional IP Network Vulnerabilities in a Flat Environment

Known vulnerabilities and attacks are migrated from the Internet and TCP/IP LANs to the TCP/IP mobile networks. Because all network equipment is exposed to unauthorized access attempts, it is critical that operating systems and platforms undertake a hardening process before being deployed into production environments. Mobile devices will also be targeted by attackers and vulnerable ports can be exploited by intruders.

Mobile networks present a flat environment without segregation. An attacker on the mobile network can target mobile devices and the mobile network infrastructure.

Flora and Fauna

Computer worms, viruses, and spyware threaten the mobile network and can cause significant damage if internal systems are infected. If mobile devices are infected with malware, they can send packets in an attempt to infect other devices on the network. This traffic can cause a service outage and a general degradation of service. Large scale infections on email systems ("love you virus"), web servers ("red worm," "Chinese worm"), and databases ("Slammer") have proven that a simple attack aimed at systems with a common vulnerability can reach hundreds of thousands of victims in a matter of hours.

Attackers do not need deep knowledge of how mobile networks operate in order to cause significant damage. TCP/IP-based worms with simple payloads can dramatically affect the availability of service.

Flood the Gates

The TCP/IP protocol and stack implementations have a number of vulnerabilities that can be exploited by attackers. A simple example is the "SYN flood attack," where large number of packets sent by a single mobile device can crash connected systems. Other attacks use spoofed IP addresses to cause response floods from multiple mobile devices to a central server. Additionally, many TCP/IP stacks contain vulnerabilities that can be exploited to crash vulnerable elements on the network and dramatically increase bandwidth consumption.

TCP/IP attacks can be started from the Internet or from within the mobile environment from anywhere in the world. This makes it even harder to stop attackers who are outside the reach of local law enforcement.

MAN-in-the-Middle

TCP/IP communications, and even some implementations of SSL, include vulnerabilities that allow attackers to compromise private communications by capturing the initial handshake between communicating parties and applying a man-in-the-middle attack. The attacker can eavesdrop on the communicating parties conversation, capture all information exchanged, or relay false messages between the parties.

Denial of Service

Denials of service (DoS) and distributed denial of service (DDoS) attacks usually involve overwhelming the target site with external communications requests that consume all its resources. DoS attacks can have a particularly serious impact on mobile networks that are part of the critical infrastructure in developed countries. Individuals and businesses that rely on mobile networks for their day-to-day functioning are severely impacted by any widespread blackouts. And these attacks, if successful, have a direct impact on the image and revenue of the target victim's service providers.

Air Time

The radio spectrum is a scarce resource. Simple attacks aimed at overloading the available spectrum can have a high impact on service and can result in a denial of service attack. Carefully prepared traffic originated from the Internet can be targeted at a number of base stations, ultimately depleting the available wireless spectrum and affecting service levels.

Mobile networks are based on the concept of shared access to wireless channels and air resources (RF bandwidth). This principle allows mobile devices and radio network controllers to become dormant until required. With dormant connections, numerous mobile devices are able to share available resources, minimizing the power consumption and extending the device's battery life.

Attackers sending frequent packets with intervals shorter than the dormancy timeout can cause mobile devices receiving these arbitrary packets to initiate new connections, consuming air resources from the radio network controller. If enough mobile devices are made to maintain active sessions, air resources will be depleted and valid subscribers will be unable to connect.

Joining Forces

Another variety of DOS attack that applies to the mobile environment involves flood attacks aimed at the base transceiver station. Attackers can start low volume attacks, tuned to increase the number of channels consumed by open connections and cause valid subscribers to lose access to the service. Attacks can be launched from inside the mobile network, or from the Internet, with a frequency just below the dormancy threshold defined by the mobile operator. Connections are then kept alive by the base transceiver station.

A variation of this attack includes port scans of mobile devices, computer worms scanning for other vulnerable systems, and in some cases misconfigured settings for VPN keep-alive-traffic and heartbeats.

Keeps on Going

Attackers can easily attack mobile environments by sending frequent arbitrary packets to high numbers of mobile devices. This ensures that these packets arrive at a specific interval within the dormancy timeout thereby forcing the mobile device to maintain a high power consumption until its battery power is drained. By increasing user power consumption, attackers can keep hundreds of subscribers from being able to use their phones.

Billing Fraud

In addition to pure technical attacks, intruders can take a profit-oriented approach. Attackers can hijack active subscriber IP addresses and spoof them in order to use fee-based services. Another potential threat is the dissemination of worms or viruses that target handsets and infect the system by sending SMS to high rate numbers that can be mixed between legitimate companies and fraudulent ones.

Countermeasures

To reduce the impact of attacks on the mobile environment, some of the following countermeasures can be implemented:

Increased Visibility

Until recently, mobile operators transmitted only voice traffic across their networks. New data services had some initial effect on the volume of voice minutes being used by subscribers, and some mobile operators included clauses in their data contract banning the use of the data service to transfer voice.

Mobile operators need to look inside the data traffic and analyze the information flowing across the network. They need to understand the type of applications and content being used by subscribers, and over time determine what types of service packages should be offered to the subscriber base. This analysis also helps to increase the security of the network by facilitating the detection of dangerous traffic and attacks to the infrastructure.

Bandwidth Allocation

Subscribers have different access requirements at specific times during the day, forcing mobile operators to provide guaranteed bandwidth allocation in alignment with specific subscriber profiles. Bandwidth allocation ensures that the customer experience is consistent with the access purchased by the subscriber.

With bandwidth allocation the mobile infrastructure can be protected by blocking excessive connection requests and ensuring that at all times the infrastructure is within expected and designed bandwidth usage limits.

User Quotas

Always on connections have the potential of disrupting service quality to other subscribers. Usage quotas can be deployed within the mobile network to ensure users are following acceptable use policy. This approach avoids the possibility of an always on connection consuming all or part of the available resources.

Peer-to-peer networks are another demand on the mobile network. These networks account for high bandwidth consumption and impact the experience of other subscribers. By limiting overall data consumption for each user, the amount of consumption is managed against needs and service agreements.

The data usage quota protects against long-term service degradation. For example, a subscriber may decide to use his allocated quota in a short period of time, for example, early in the month. Assuming a normal distribution of others subscribers doing the same thing, resources are freed up for use later in the month. This allows the operator to better balance all subscriber usage.

Application Management

Filters can be deployed to enhance or reduce the bandwidth available to specific applications. Mobile operators can deploy specific filters to ensure that IPTV and in-house services have the maximum bandwidth available. At the same time, other applications, such as peer-to-peer and VoIP services, can be controlled to limit their bandwidth availability to predetermined values – again balancing overall use of network resources.

Known viruses and worms can be filtered by blocking connection requests to known vulnerable applications. Specific packet payloads targeted to vulnerable applications can be blocked.

Signalling Attacks in Mobile Networks

The security of computing systems is based on three basic principles: confidentiality, integrity and availability. System availability of networks and services can be significantly impaired by Denial of Service (DoS) attacks which can take various forms which differ according to the technology being considered. Thus DoS attacks for IP (Internet Protocol) networks differ significantly from DoS attacks against mobile networks.

Mobile networks are susceptible to DoS attacks, mostly because of the networks' openness to the Internet, the use of deterministic procedures, and the use of basic design principles based on "typical" user behaviour. There were huge advances from an algorithmic, manufacturing 'and software perspective, pushing forward the innovation of mobile smart devices and applications, which operate over a mobile network - while the network itself did not keep up with the pace. One of the problems caused by these circumstances is the appearance of DoS attacks known as signalling storms, which overload the control plane of the mobile network, unlike many previously known data plane flooding attacks.

Network security is ranked as one of the top priorities for future self-aware networks, which is why there is well established research in the field. Furthermore, while work in focuses on a general defensive approach against DoS attacks in future networks, signalling storm specific research can roughly be categorised in the following groups: problem definition and attacks classification; measurements in real operating networks; modelling and simulation; impact of attacks on energy consumption; attacks detection and mitigation, using counters, change-point detection techniques,

IP packet analysis, randomisation in RRC's functions, software changes in the mobile terminal, monitoring terminal's bandwidth usage, and detection using techniques from Artificial Intelligence.

The communication schemes may be opportunistic and attacks may use similar opportunistic means to access IoT devices, viruses and worms will continue being important threats and they can diffuse opportunistically through a network, video input is one of the uses of the IoT and video encoding can also be specifically targeted by attacks. Furthermore, many network services are organised to flow over overlay networks that cooperate with the Cloud to offer easy deployable and flexible services for the mobile network control plane. Thus research needs to remain alert to such developments.

Network Model

The proposed model describes a general network architecture, focusing on its radio access part, from the perspective of both, the control and data (user) plane. It's envisioned to represent different mobile network technologies, which is achieved through representing the resource allocation in the data plane as a "black box" where different technologies' sub-models can be plugged in, while keeping the control plane unchanged. The core part of the model consists only the basic elements of the architecture, such as multiple Base Station (BS) nodes connected to a single network controller consisting one Signalling Server (SS) node, and the communication stage nodes.

An example workflow captured by our model goes as follows. When a mobile terminal wants to communicate, it sends a connection setup request through the control plane of the network, which needs to be processed at the BS and the SS. If admitted, the mobile proceeds to communicate in the data plane of the network, in sessions (each comprising multiple data packets), If a call is blocked, then the mobile may either leave the network or attempt to reconnect with a probability that depends on the type of call. There are two types of calls or connection setup requests in the network: (i) normal calls representing traffic from legitimate users or applications, and (ii) attack traffic generated by malicious or malfunctioning applications that may overload the network. The network model is open with calls joining and leaving the network, representing for example the arrival and departure of mobiles to WiFi areas. Its parameters are defined in table where the superscript $r \in \{n, a\}$ denotes the class of a call (normal n or attack a).

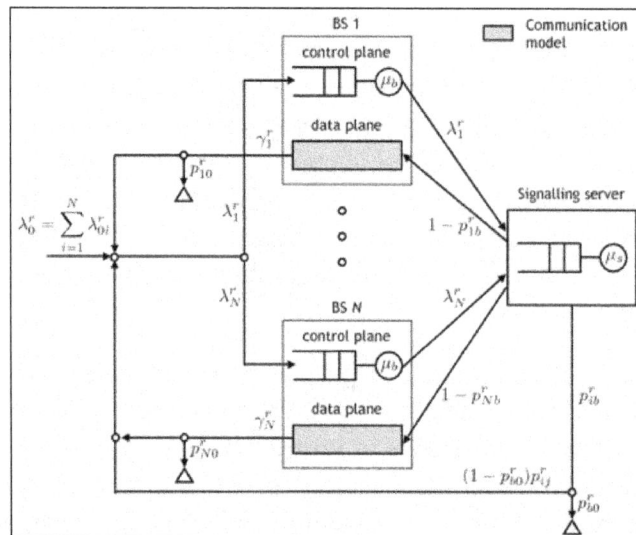

A model of the radio access part of a mobile network.

The Main Parameters of the Model

- N : Number of cells covered by one signalling server.

- λ_{0i}^r : Rate of new class-r calls joining cell $i \in \{1, ..., N\}$, which corresponds to mobile phone activations and handovers by roaming users.

- λ_i^r : Rate of class-r connection requests traversing the i-th BS. These include calls joining from outside the network, calls that have been successfully served and return as new calls, and calls that retry connecting after not being admitted at cell j due to insufficient data channels.

- λ_s^r : Total rate of class-r calls arriving at the SS, $\lambda_s^r = \sum_{i=1}^N \lambda_i^r$.

- γ_i^r : Rate of class-r calls that timed out after being admitted to cell i.

- p_{ib}^r : Proportion of class-r calls not admitted for communication at cell i.

- p_{ib}^r : Probability that a blocked class-r call leaves the network; p_{b0}^a represents attackers' stubbornness while p_{b0}^n reflects human persistence.

- p_{b0}^r : Proportion of class-r calls leaving the network after successful service at cell i.

- p_{ij}^r : Proportion of class-r calls joining cell j after being blocked at cell i given that they stay in the network, i.e. $\sum_{j=1}^N p_{ij} = 1$.

- μ_b : Class-independent service rate of connection requests in the BS, representing the cell signalling capacity.

- μ_s : Class-independent service rate of connection requests in the SS, representing the SS capacity.

- t_0^r : Inactivity timer.

We assume calls arrive from outside the network according to independent Poisson processes and the service times in each node are independent and exponentially distributed. Since calls may be blocked at the SS due to congestion, the aggregate arrival processes at different parts of the network are not Poisson. Nevertheless, to simplify matters so as to obtain analytical solutions, we make the approximation that all flows within the network are Poisson. The service time distribution for the BS and SS nodes in the signalling stage is same for both classes of calls, because the signalling procedure undertaken by the network does not distinguish call classes. On the other hand, in the communication stage, the service time distribution is distinct for different classes of calls because of the different bandwidth usage behaviour of the normal and malicious calls.

The flow of calls in the above model could be expressed in a closed form as follows. The total arrival rate of class-r connection requests at BS i is the sum of the rates of (i) new calls, (ii) returning calls that timed out, and (iii) calls that were blocked at a cell j by the SS and are attempting to connect at cell i:

$$\lambda_i^r = \underbrace{\lambda_{0i}^r}_{\text{new calls}} + \underbrace{\gamma_i^r(1-p_{i0}^r)}_{\substack{\text{reconnecting after}\\\text{timeout}}} + \underbrace{\sum_{j=1}^N \lambda_j^r p_{jb}^r (1-p_{b0}^r)p_{ji}^r}_{\substack{\text{joining after being blocked}\\\text{at cell } j \text{ due to congestion}}},$$

where the proportion of blocked calls P_{ib}^r and the rate of admitted calls that has timed out $\tilde{a}_i^r$ depend on λ_j^r, $\forall_j$. The model as presented is suitable for modelling different mobile technologies under an attack.

User Behaviour Model

An important part of the network model is the user behaviour model. In general, the two classes of calls have different service time distributions. A normal call, for example web browsing traffic, would usually happen in bursts which would occupy the channel for a longer period. Contrary, attack calls would usually transfer only a small portion of data in order to trigger quick bandwidth allocations and deallocations. The two patterns are depicted on Figure with T^n denoting the normal session duration and T^a the attack session duration, and s and q respectively denoting "service" and "quiet" periods. we need to estimate the average session duration $E[T^r] = 1/\mu^r$.

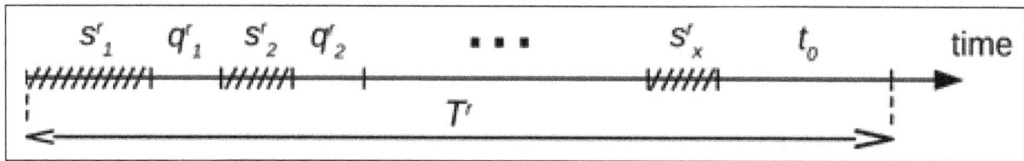

The user behaviour model describing the duration of a single data session T^r of class r.

Figure could be translated to a Markov Chain model as in Figure, using the states: service (S), quiet (Q), and end of session (F). The transitions among S and Q states are controlled with α^r, and β^r, where $1/\alpha^r$ is the average communication time of a class-r burst, and $1/\beta^r$ is the average duration of a quiet (inactivity) period, regarding class-r calls. The timeout rate is given with $\tau = 1/t_0$.

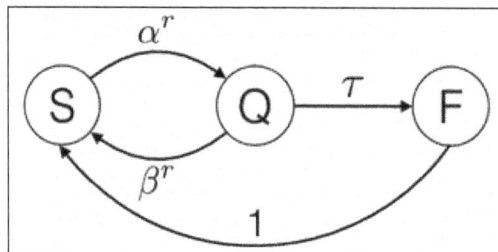

State diagram of the user behaviour model.

Let us denote with Π_i the probability of the session being in one of the states $\{S, Q, F\}$. The average session duration could be found using the following ratio:

$$\frac{\Pi_S + \Pi_Q + \Pi_F}{1 + E[T^r]} = \Pi_F.$$

Solving the balance equations yields the state probabilities in equilibrium, and the above equation solves to:

$$(\mu^r)^{-1} = E[T^r] = \frac{1}{\mu^r} = \frac{1}{\alpha^r} + \frac{1}{\tau} + \frac{\beta^r}{\alpha^r \tau}.$$

In the above expression, one can see that when the timeout is very short, with $\tau \to \infty$, the average session duration tends to the communication time of a single burst $1/\alpha^r$. Modifying the α^r and β^r parameters, this modelling approach can be used to investigate different traffic types, and different attack patterns.

Detection and Mitigation

Counter Detection

The Counter detection mechanism enables detection of signalling storms per mobile terminal in real-time. It is based on counting the repetitive bandwidth allocations of same channel type (eg. a shared FACH or dedicated DCH channel in a 3G UMTS network). It is envisioned as a lightweight mechanism that should not impose any processing, storage, and memory problems if implemented on a mobile terminal.

Decision Making

The mechanism requires two input parameters: the time instances of bandwidth allocation and the type of bandwidth allocation, which are stored in memory for the duration of a time window of length t_w. A decision of an attack being detected is simply taken when the number of repetitions reaches a predefined threshold called counter threshold - n. The length of the window t_w is chosen such that $t_w > n \cdot t_I$, where t_I the duration of the inactivity timer of the attacked state. The upper limit of t_w is set according the memory and storage capacities of the device on which it is implemented.

Evaluation

Figure shows the performance of the described detection algorithm using a ROC curve, as calculated with the SECSIM simulator. A threshold of n = 3 could be a suitable choice resulting in around 40% true positive detection P_{tp} and less than 0.2% false positive detection P_{fp}.

Bandwidth Monitoring Detection

The Bandwidth monitoring detection mechanism uses a simple idea of tracking the bandwidth usage of each mobile terminal in a given sliding time window, and calculating a cost function to estimate the likelihood of a terminal performing a signalling attack. It's based on previous analyses which showed that signalling storms are inefficient bandwidth users. The mechanism monitors two input parameters: the total time that the terminal spends while allocated bandwidth within a given time window t_w (denoted with t_D, and t_F respectively for DCH and FACH states in 3G UMTs), and the time which the mobile terminal is allocated bandwidth but does not transfer any data in a time window t_w (denoted with t_{Di} and t_{Fi}). Whenever resources are de/allocated, the

detector calculates the ratio $\frac{t_{Fi}+t_{Di}}{t_F+t_D}$, which is then rolled in time using the Exponential Weighted

Moving Average (EWMA) algorithm as:

$$C[k] = \alpha \frac{t_{Fi}[k]+t_{Di}[k]}{t_F[k]+t_D[k]} + (1-\alpha)C[k-1],$$

where $k \in \mathbb{N} > 0$ is the index of the state change, $0 \le \alpha \le 1$ is a weight parameter and $C[0] = \frac{t_{Fi}[0] + t_{Di}[0]}{t_F[0] + t_D[0]}$ is the initial cost value. As defined, C is between 0 and 1 with values closer to 1 indicating higher probability of an attack.

For decision making, we define two thresholding rules, and a rule based on the cost function. Observing the cost C, and having calculated an average C_{avg} over all historical C values, a simple rule of $C \ge \beta C_{avg}$ can be used to detect an attack. A second rule is using an upper threshold θ^+ above which we make a decision of an attack. This rule helps in detecting attacks with very small attack rate, for which the cost function rule cannot be used, because $\beta C_{avg} > 1$. A second threshold is defined as lower threshold θ^- below which we assume a normal behaviour of the mobile terminal. The θ^- rule helps in protecting mobiles with normal behaviour of high activity, which are assigned a low value of C_{avg}. Setting up these thresholds should be based on offline traffic analysis by the mobile operators.

Evaluation: The performance of the Bandwidth monitoring detection algorithm is depicted with the ROC curve on figure, which combines the P_{fp} and P_{tp} metrics. Values in the top-left corner of the graph are most desirable, as it produces the highest true positive and lowest false positive detection probabilities. The simulation results suggest that $\alpha = 0.3$ is the most suitable value, producing 95% true positive and 0.04% false positive detection.

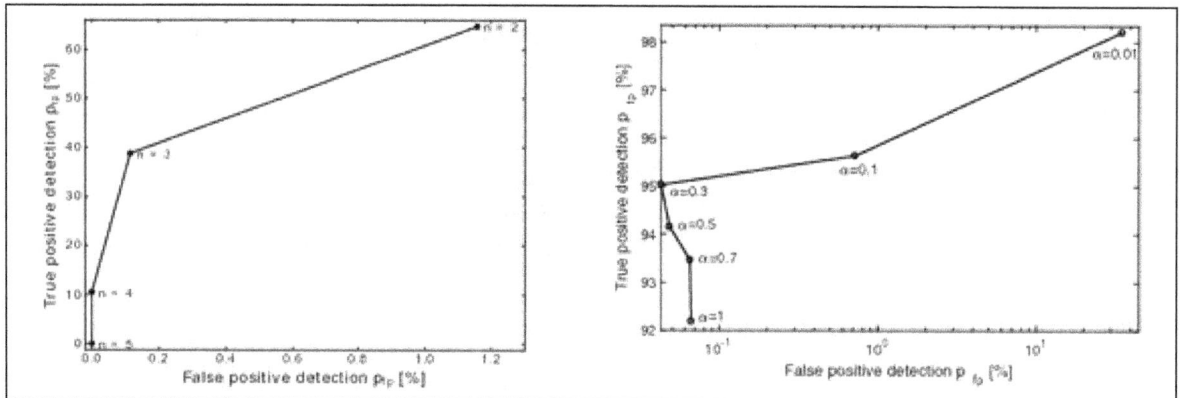

Dynamic Timer Mitigation

Mobile networks today use a fixed value for the inactivity timer with possible manual corrections for specific situations, which we consider to not be the optimal approach. While it plays an important role in controlling radio resource allocation, being a trade-off parameter between the bandwidth reuse and number of connections, this section examines if it could possibly play a similar role controlling the impact of a signalling attack on the network.

One possible approach is to increase the timer linearly with the load on the signalling server, after a signalling load threshold value θ is reached:

$$t_0(\lambda_s) = \begin{cases} t_0^{min} & \lambda_s > \theta, \\ \dfrac{(t_0^{max} - t_0^{min})}{\lambda_s^{max} - \theta} \cdot (\lambda_s - \theta) + t_0^{min} & \lambda_s > \theta, \end{cases}$$

Where λ_s^{max} is the maximum allowed load on the signalling server, θ is a load threshold and t_0^{min} and t_0^{max} are the minimum and maximum values that the timer can take. In real operating network, these parameters need to be estimated from statistical observations.

Results: Using the model, we select a data plane model with $m = 20$ non-sharing data channels, such as in 3G UMTS Rel. 99, modelled as M/M/m/m Markov chain. The rest of the parameters are selected as follows:

$$\lambda_0^n = 1, p_0^n = 0.9, p_0^a = 0.1, p_{b0}^n = 0.9, p_{b0}^a = 0.3, \lambda_e = 0.05,$$

$$t_0 = 2s(\text{static}), t_0^{max} = 60s, t_0^{min} = 2s, \lambda_s^{max} = 5\,\text{calls/s}\,\theta = 3\,\text{calls/s}.$$

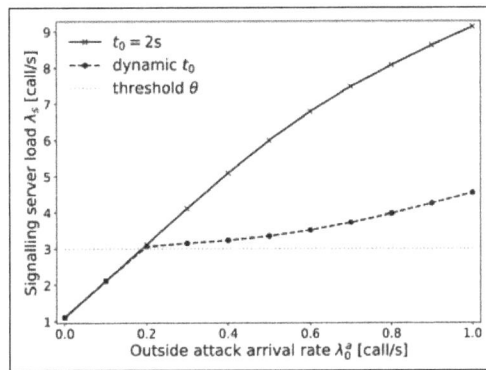

Signalling server load for static and dynamic inactivity timer.

Figure shows the comparison of a static and dynamic inactivity timer for varying network load. The dynamic timer activates when the threshold load θ is reached and manages to lower the resulting network load, compared to the static approach. Although the timer can play a control role, it cannot completely mitigate a signalling storm. One downside of using this approach is increasing the portion of normal calls that don't get a service. Therefore, the timer controls the trade-off between the signalling load in the network and the number of unserviced normal calls.

Attacks on LTE in 4G

4G is the 4th generation technology. In 4G, more bandwidth and services are provided as in comparison of 3G. LTE and Wimax are two technologies that increase the levels of 4G.

Wi-Fi Security

Mainly this arises due to the change in the format of words during coding and decoding. After finding its weakness, we can approach, towards an efficient system.

Wi-Max Security

When the system does not support the management and integrity of the system, then this problem arises. After careful observation, its security system can be raise in a better way.

3GPP LTE Security

It is a standard form used for wireless communication. Different reasons, that effects it, one of the major reason is not to synchronize in a proper manner and security issue arise.

Different Attacks in Long Term Evolution

Attacks affect the integrity of the system. There are mainly two types of attacks, one is active attack and other is passive attack. When the attackers only aim is to take the information and then it is passive attack. But its aim is not only take the information, but effect the integrity of the system is active attack. Passive attack is such that traffic problem arise, during communication. Other is unauthorized user, get the information is eavesdropping. Passive attacks are such as denial of service attack, resource consumption, masquerade attack, replay attack, information disclosure; message modification etc. Sensor network is used in wireless communication on a wide level. Larger number of nodes relates to the sensor networks, due to which different problem arise. Denial of service attackers, aim is to target the destination information, take the information. So, for, its need higher security, because on wireless sensor can be easily attacked.

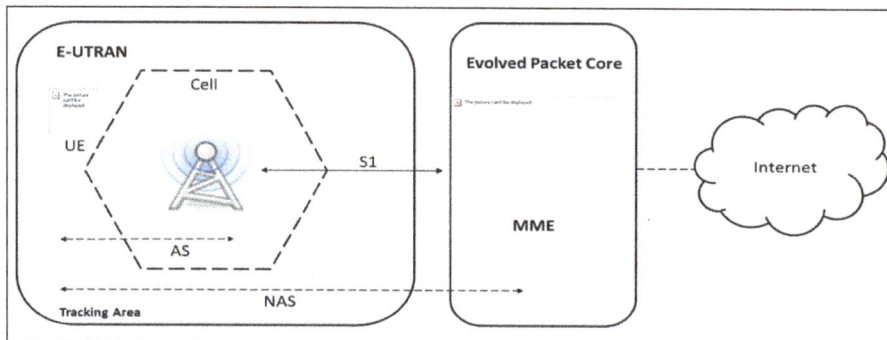

Different security issue arises in wireless 4G network due to threats and attacks. These attacks on physical layer and MAC layer attacks in 4G. These issues affect the integrity of the system. Attacker targets the physical layer or the MAC layer.

Physical Layer Issue

A private key is used between the two users when both the users do not have the private key then physical layer is used there, to convey the information between them and its need to discuss the different security issue in physical layer.

Wimax and LTE have two keys at the physical layer. These are interference and scrambling attacks. Interference is considered due to the noise and multicarrier noise such as white Gaussian. In multicarrier interference, attacks recognized the carrier used by the system and inject a narrowband signal, into this carrier.

Scrambling is also an interference targeted by an attack. An attacker targets on management.

Mac Layer Issue

There are the different issues arise on the Mac layer. These are location tracing, when unauthorized user traced the location, and get the information. Denial of service is the second, between them when the bandwidth is stealing; then, bandwidth stealing is another security issue arises. Other issues related to its architecture also arise. These are the, mainly reason, the security issues on Mac layers. Wi-max and LTE are the major standard here.

There are different security issues arises on the physical and Mac layer of the LTE standard. Discuss both, physical and Mac layer of the system. LTE security verifies by authentication, integrity and encryption. In the first verify UE's identity by challenging the UT use the key and report a result after checking it. Femtocells generally have low power and lower rate of data transmission. So that it can be used by the user. Long term evolution standard is used to meet with its demands. But different challenges arise due to the use of this. Those challenges affect the integrity of the system. One of them is interface of the air between the mobile device and the femtocell. Attackers attack on this air interface. Other attack is on the femtocell itself. And, third one is attacked on the core network. These are the attacks, affected the integrity of the system.

IP Spoofing in Mobile Network

Mobile communication started as voice service with focus on AMPS(Advanced Mobile Phone Service), the representative 1G (First Generation) mobile communication, in 1978, and voice and data service began to be provided at the same time with 2G represented by CDMA (Code Division Multiple Access). Afterwards, mobile communication is evolving beyond 3G WCDMA (Wideband Code Division Multiple Access) capable of providing faster data services, and LTE (Long Term Evolution) called 3.9G into 4G mobile communication. Early mobile communication service was developed for voice communication, but as the Internet in the wired environment advanced, demands for mobile service, which provides mobility based on mobile communication service, increased. Accordingly, data networks for providing data communication as well as voice service were added to mobile networks, and voice is processed as VoIP (Voice over Internet Protocol) in accordance with the All-IP communication paradigm. The importance of IP-based data networks is growing gradually.

The data service provided by early mobile networks started out as a type of mobile service provided by communication companies in a limited way, but the advances of the Internet and mobile operating system created a mobile ecosystem. At the same time, mobile networks are open to the Internet, and various services in the wired environment were offered in the mobile environment as well. As a result, the data communication volume of mobile networks increased explosively, and is expected to rise continuously in the future.

As the increased traffic includes not only the traffic for various mobile service, but also the traffic in the wired environment that could not be seen in existing mobile networks, unnecessary abnormal traffic also increased. In the conventional wired environment, the increased traffic did not mean much to the receiver unless there are large quantities of abnormal traffic like UDP (User Datagram Protocol) packets and TCP (Transmission Control Protocol) packets, which failed to connect. However, in mobile networks, due to the narrow bandwidth, complicated signaling for management of wireless resources, and operation of limited resources, traffic, which did not matter in the existing wired environment, can become a security threat in the mobile network. Also, aggressive security threats, likely to cause the failure of mobile networks, may cause not only data services, but also voice services to fail unless they are responded to in advance.

At present, as most security systems are optimized to IP-based wired networks, they processes mostly IP protocols, and identify send and receive objects based on IPs. However, mobile networks protocols specialized for mobile networks like GTP (General Packet Radio Service Tunneling Protocol, and IP is not the unique value that can identify a user. Also, as abnormal traffic for mobile networks may look different than that for the wired environment, it is very difficult to bring the security systems for the wired environment inside the mobile network.

The mobile network is the backbone network of the country. If an important infrastructure fails, the repercussions are enormous.

Mobile Network Attacks using IP Spoofing

IP Spoofing means that the sender alters an IP address other than assigned to the sender as the source IP. IP Spoofing makes it difficult to trace the IP of the attacker, and has been used by various attacking techniques like DoS attacks in the wired environment. But IP Spoofing can be filtered in a limited way by the Network Ingress Filtering of the switch or router in the wired environment.

As IP Spoofing is not taken seriously in the mobile environment, the resulting security threats were not taken into consideration in a big way, but IP Spoofing in the mobile environment can lead to overbilling and power consumption for certain UE, occupy the wireless resources of the mobile network, and induce abnormal traffic into components in the mobile network such as GGSN, P-GW.

Mobile networks may be configured differently depending on service providers, but mostly, as shown in figure (a), they have NAT (Network address translation) which provides communication between UEs and external services. Unless the internal network requests communication, NAT cannot allow attempt to communicate with an object on the network from outside. This has something to do with the distinct characteristics of the mobile network. As the communication packet in the mobile network is to be billed, unnecessary traffic must be minimized, and only required communication for service must be available. And the mobile network must go through a complex signaling process to send data to the UE. In particular, as the wireless resources between the base stations and the UEs are limited for each base station, they are allocated to the UE only when necessary, and then released. Likewise, the UE with limited electric power is not active all the time. Instead, it is deactivated after a certain amount of time to reduce the power consumption of the UE, and wireless resources are released. If IP Spoofing is used, however, NAT will be incapacitated, and abnormal traffic may be brought to the IP that did not attempt communication inside the mobile network, as shown in figure (b).

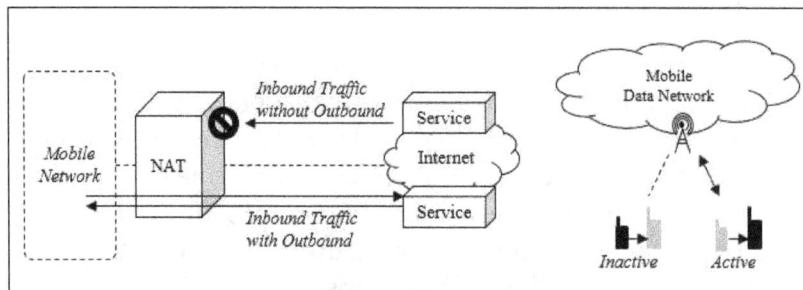

(a) NAT structure, (b) status change of UE.

IP Spoofing in Mobile Data Network

In the mobile network, the IP packet generated at the UE is relayed to GGSN or P-GW. It means that the IP packet of the UE passes the packet data network unconditionally in any case, and through additional security devices like NAT. At this time, NAT maps the source IP/port of the outbound packet and the external destination IP/port, and enables communication with the IP of NAT by creating the NAT table.

Based on the above structure, the conceptual procedure of IP Spoofing in the mobile network is illustrated in figure. Before IP Spoofing is described, it will be assumed that there are UE A and B, and both UEs are capable of data communication. And 192.168.1.1 was assigned to UE A, and 192.168.2.2 was assigned to UE B. And it will be assumed that the external service server provides echo service at 9.9.9.9/456, as shown in figure.

- Step 1: First of all, UE A alters the source IP to 192.168.2.2, not its own IP 192.168.1.1, and sends it to the external service server (9.9.9.9/456). The spoofed packets sent by UE A pass through a data network like CN and are transmitted to NAT.

- Step 2: NAT maps the source IP of the IP packet 192.168.2.2 and the external IP 9.9.9.9 to create the NAT table, and modifies the source IP (port) with the IP of NAT, and transmits the packet to the external service server.

- Step 3: The external service server sends response packets to the source IP of the received IP packet, and through NAT to CN. CN and EPC search for the UE corresponding to 192.168.2.2 which is the destination IP address, verify that it is UE B, establish the GTP tunnel of UE B, and send response packets.

- Step 4: UE B receives unwanted response packets.

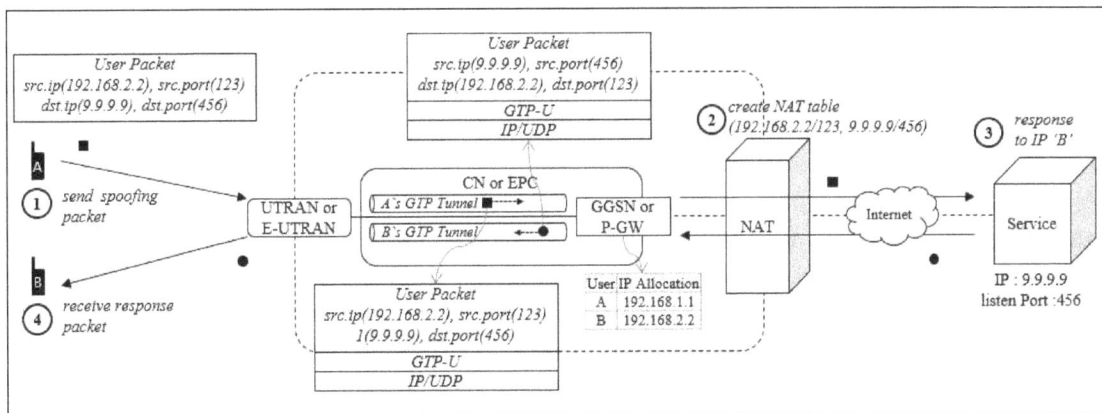

IP spoofing in mobile data network.

Attack using IP Spoofing

The core of IP Spoofing is that unwanted abnormal traffic, not the requested communication traffic, can be brought to unspecified individuals in the mobile network. IP Spoofing can be used to send an unwanted bill to users with a certain IP, and sends abnormal signals in the mobile network concurrently to cause overload in the network or consume the resources of the UE and mobile network. Also, it may cause overload in important components in the mobile network.

Overbilling for users In general, charging for data usage in the mobile network is done on the basis of IP. For example, a component like GGSN which can handle user packets, outputs information on traffic use based on IP and the charging system gets such information together, and calculates data usage. Of course, some traffic caused by well-known worms and certain DNS service may be excluded from the billing according to the policy of the service provider. If the type of the traffic like the number of send & receive packets, can be viewed as normal communication, however, charging may be done for the IP.

As shown in figure, attackers will execute abnormal service in a certain external server. If packets are inputted, this service sends a large volume of traffic to the packet source. Attacker A sends the packets, spoofed to a certain IP (B's IP), to a pre-specified external server. The external server sends a large volume of traffic to the source IP (B's IP) of the received packets, and user B receives a large volume of unintended traffic. In this process, for example, attacker A sent less than 2bytes of packets for the attack command, but user B can receive a large volume of packets bigger than 1Kbyte as specified in the external server as long as MTU allows it. In this case, attacker A is billed for 2bytes on the surface, and user B is billed for the received packets. Billing details may vary depending on the billing policy of the mobile communication service provider, but what is important is that overbillings can be made due to IP Spoofing.

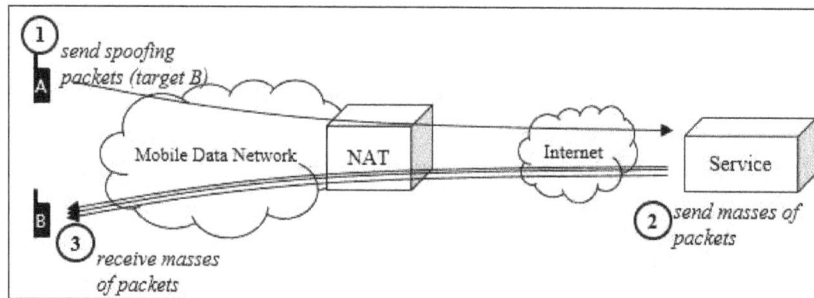

Overbilling attack.

Consumption of resources: Wireless resources in the mobile network are limited resources and the complicated control process is essential for establishing and managing such resources. For example, the UE has the Active mode and the Inactive mode. The UE requests and sets up wireless resources in the Active mode only when necessary, and then goes into the Inactive mode to save unnecessary resources, and hands over wireless resources so that other UEs can use them. Also, when the base station receives communication requests from outside, as the UE is already using wireless resources if it is in the Active mode, it will simply transmit data, but if the UE is in the Inactive mode, the paging signal for looking for a UE will be sent to the base station wirelessly, and the wireless resources for data communication with the responding UE will be set up. If the UE's Active/Inactive switching time is t, and communication traffic arrives every $t + \alpha$, the signal and wireless resource setup/release procedure for repeated wireless resources setup and the switching of the UE's Active/Inactive mode may be repeated. Or if communication traffic comes every $t - \alpha$, resources may be occupied continuously, and other UEs use of wireless resources may be hindered. If these attacks are not happening to one UE, but to UEs in a certain area, or on a large scale, it may be a great burden on the mobile network.

The attacker implements a simple echo service for periodically sending packets to an external server, and as illustrated in figure, spoofs packets to multiple IPs, creating traffic small in size, but aimed at multiple targets. Then, response packets for multiple IPs are generated externally, and sent to respective UEs. If the UE is Active, it will continue to occupy wireless resources, whereas if it is Inactive, it will switch to Active and masses of signals for setting up wireless will be generated concurrently, causing DoS attacks in the network.

Overload for mobile network: These expensive components such as SGSN, GGSN, S-GW and P-GW, process the large volume of traffic in the mobile network. So Availability is important, but the mobile network has been regarded until now as a safe network regarding which there is no concern over abnormal traffic sent to the machines from outside or inside. The attacker implements an attack service that sends abnormal traffic to an external server, and as shown in figure, the UE sends the spoofed packets to the IP of the internal machine. The external attack server can send a large volume of attack traffic to the internal network machines, and the internal machines receiving the attack traffic may find it burdensome to process the traffic. For example, if the attacker acquires the IP of the internal GTP-related machine through GTP Scan, and the attack server abnormally alters the GTP protocol, an important processing protocol, and sends it as attack traffic, each key machine can receive DoS attacks due to unnecessary processes like exception handling of abnormal GTP packets and errors.

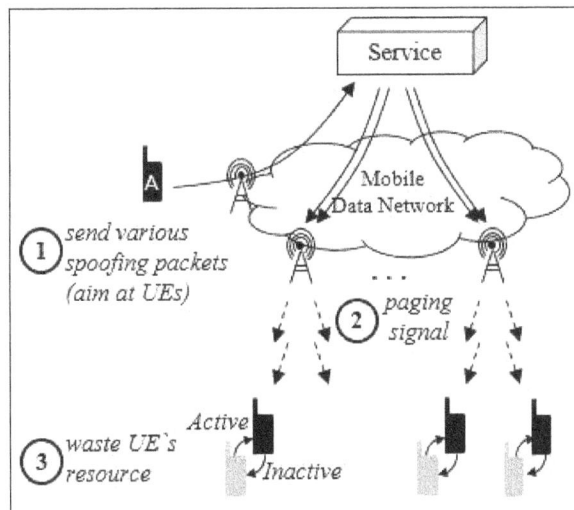

Resource consumption attack.

Attack in Real Field

To verify a security threat by IP spoofing in the commercial mobile network, we conducted the test with an Android-based smart phone that is relatively free to manage and control. For test, the attacker and the victim used the same UE, i.e. Samsung Galaxy Nexus, subscribing to a flat rate monthly pricing on a commercial 3G service network. The attacker's UE was rooted to transmit IP Spoofing packets, and the Linux shell was loaded so that the IP Spoofing program could work. Wi-fi was deactivated in each UE, and the UEs were connected to the 3G network. Apps like Myip were used to check the IP of the UE, and both the attacker and the victim used tcpdump to collect packets from inside the UE. The packets spoofed to the IP of the victim's UE from the attack UE were sent to the external echo server. We checked the network traffic collected from each UE, and found that, as shown in figure, the spoofed packets sent by the attacker were sent to the external server, and the packets sent by the external server were received by the victim's UE. The receiving UE generated

ICMP for the received packets. (The IP of the UE is a public IP band on the surface, but it is used only inside the mobile network, and is changed through NAT for Internet communication).

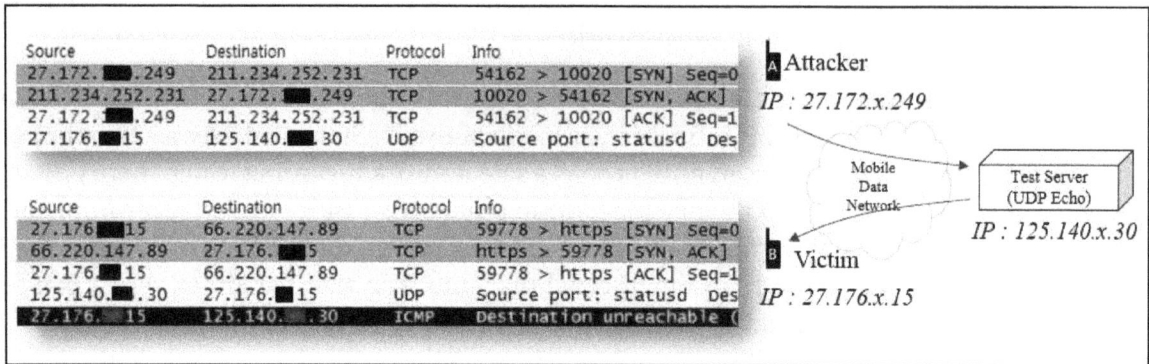

IP spoofing result in real field.

Overbilling Attack

In the experiments for checking the occurrence of overbillings due to IP Spoofing, the active applications in the UE were suspended as much as possible in advance, and the current data usage was checked. Then, we used apps like Myip to check the IP of the victim's UE, and sent the UDP packets, spoofed from the attacker's UE to the IP of the victim's UE, to the external attack server. After the attack is completed, we checked the charging status of each UE, and obtained the results as shown in table. The results show that the attacker can cause overbillings to a user with a specific IP.

Table: Result of Overbilling Attack.

Option	Scenario 1	Scenario 2
Size of attack packet	2byte	2byte
Number of attack packet	10,000	110,000
Traffic from Attacker	0.02MByte	0.2MByte
Traffic from Attack Server	9.9Mbyte	109.3Mbyte
Receive traffic at Victim	9.8MByte	106.7MByte
Receive traffic at Attacker	0byte	0byte
Billing for Victim (Data Usage)	9.8MByte	106.7MByte
Billing for Attacker (Data Usage)	0byte	0byte

Battery Depletion Attack

Regardless of whether there was any charge made to the UE, IP Spoofing can continuously deplete the battery of the UE. We created an environment similar to that causing overbilling. And the attacker sent the attack message when the battery of the victim's UE was fully charged, and we made sure that the external attack server sent an unlimited volume of traffic. As a result, as illustrated in figure, more than 80% of the battery was depleted in just 5 hours. The battery was discharged about 12 times faster than the normal discharging speed, as shown in figure. The victim's phone was Google's reference phone(Galaxy Nexus) manufactured by Samsung at the end of

2011. According to the official specs, the smart phone had 210 hours of waiting time, and 18 hours of continuous talking time.

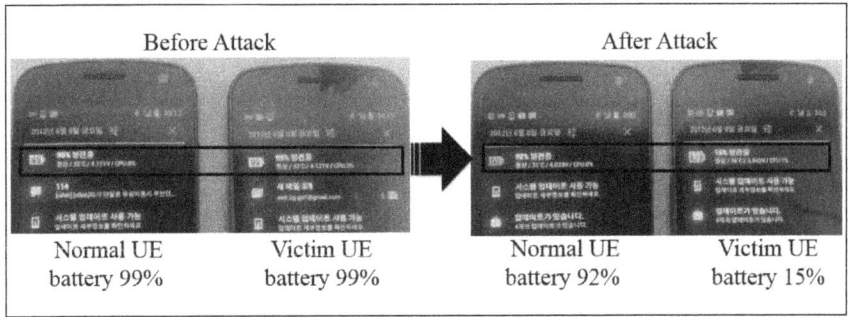

Result of battery depletion attack.

Approach for IP Spoofing Prevention in Mobile Network

In the mobile network, it is advantageous to detect user packets before getting out of the GTP tunnel in order to detect IP Spoofing. IP Spoofing is hard to detect using method like Network Ingress Filtering because the more traffic goes outside the traffic is becoming increasingly intensive. And also the L2 layer carrying user IP packets are newly created in GGSN or P-GW. Accordingly, to detect IP Spoofing in the mobile network, we must check if the source IP of the user packets transmitted by the Outbound GTP-U packet in the GTP tunnel section is valid. The process of generating the GTP tunnel using GTP-C v1 as an example is roughly illustrated in figure.

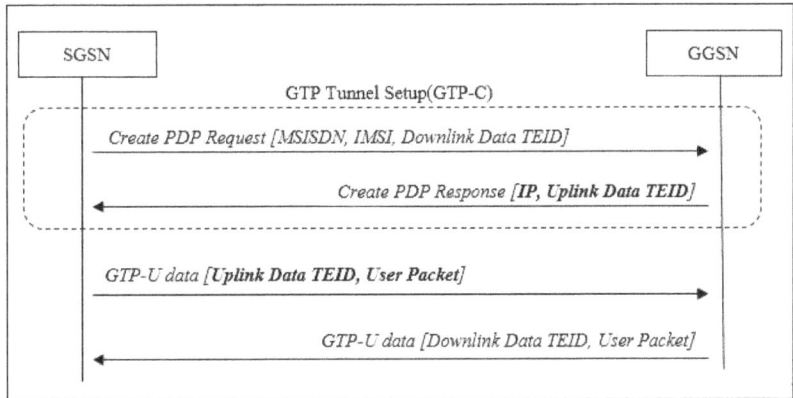

Information exchange during GTP tunnel setup.

To create the GTP tunnel, SGSN use the UE information (MSISDN, IMSI, etc.) regarding which UE the tunnel is for, and assign the IP the UE will use in response. TEID exists separately for each UE depending on directionality and packet type (GTP-C/GTP-U). When creation of a tunnel is requested, TEID (Downlink Data TEID) for sending data to the UE will be sent as well, and the response will include TEID (Uplink Data TEID) used for the data the UE sends outside. Accordingly, while the GTP tunnel is generated, depending on the direction in which the UE sends data, TEID to be used and the IP to be used by the UE will be determined. Accordingly, we can use the GTP tunnel information to detect IP Spoofing as shown in the figure. As the GTP tunnel information necessary for IP Spoofing is information that can be checked regardless of the GTP-C version, it can be applied not only to 3G, but also to GTP-C v2 used in LTE.

Mobile Device Security

Mobile device security is the full protection of data on portable devices and the network connected to the devices. Common portable devices within a network include smartphones, tablets, and personal computers.

Why is Mobile Device Security so Important?

Nowadays, over 50 percent of business PCs are mobile, and the increase in Internet of Things (IoT) devices poses new challenges to network security. Consequently, IT must adapt its approach to security. A network security plan must account for all of the different locations and uses that employees demand of the company network, but you can take some simple steps to improve your mobile device security.

Securing mobile devices requires a unified and multilayered approach. While there are core components to mobile device security, every approach may be slightly different. For optimum security, you need to find the approach that best fits your network.

Components of Mobile Device Security

Here are some solutions that can help keep your mobile devices more secure:

- Endpoint security: As organizations embrace flexible and mobile workforces, they must deploy networks that allow remote access. Endpoint security solutions protect corporations by monitoring the files and processes on every mobile device that accesses a network. By constantly scanning for malicious behavior, endpoint security can identify threats early on. When they find malicious behavior, endpoint solutions quickly alert security teams, so threats are removed before they can do any damage.

- VPN: A virtual private network, or VPN, is an encrypted connection over the Internet from a device to a network. The encrypted connection helps ensure that sensitive data is safely transmitted. It prevents unauthorized people from eavesdropping on the traffic and allows the user to conduct remote work safely.

- Secure web gateway: Secure web gateways provide powerful, overarching cloud security. Because 70 percent of attacks are distinct to the organization, businesses need cloud security that identifies previously used attacks before they are launched. Cloud security can operate at the DNS and IP layers to defend against phishing, malware, and ransomware earlier. By integrating security with the cloud, you can identify an attack on one location and immediately prevent it at other branches.

- Email security: Email is both the most important business communication tool and the leading attack vector for security breaches. In fact, according to the latest Cisco Midyear Cybersecurity Report, email is the primary tool for attackers spreading ransomware and other malware. Proper email security includes advanced threat protection capabilities that detect, block, and remediate threats faster; prevent data loss; and secure important information in transit with end-to-end encryption.

- Cloud access security broker: Your network must secure where and how your employees work, including in the cloud. You will need a cloud access security broker (CASB), a tool that functions as a gateway between on-premises infrastructure and cloud applications (Salesforce, Dropbox, etc.). A CASB identifies malicious cloud-based applications and protects against breaches with a cloud data loss prevention (DLP) engine.

Security Threat and Countermeasure on 3G Network

Existing Wi-Fi mobility is relatively limited, but the 3G network with high mobility than a Wi-Fi environment was provided. And smartphones and mobile devices as a platform for change in general (android, iOS) and the popular dissemination has done, it was possible based on a variety of mobile services.

Unlike traditional wired infrastructure, mobile networks has limited radio resources and signaling procedures for complex radio resource management. So these traffic is not a problem in wired networks but mobile networks, it can be a threat. If provide mobile services that moved from a traditional wired network services (messenger, etc.) are required to be connected anytime, and that can be an inefficient waste of limited radio resources. In addition, a narrow bandwidth of conventional wired infrastructure was not a significant problem, unnecessary traffic (scanning traffic and malicious traffic, etc.) a waste of resources, it can interfere with other users.

Recent connection between heterogeneous networks (mobile network and wired network) is sharing mutual security threats. Especially compared to the existing wired network, mobile network security for various abnormal traffic technologies was not ready. Mobile networks as a communications facility is viewed as a national infrastructure, because if it can be backed up with appropriate security technologies by hackers can be a victim of cyber terrorism, economic and social loss for mobile operators will be greater. We analyze the security threats in mobile networks and provide direction to solve it.

Introduction of 3G Networks

Structure of 3G Network

In figure, 3G network, which consists of two main elements, circuit network (CN) for voice communications and packet network (PN) for data communications. And there are RAN that related to wireless connectivity, and the system for authentication and billing functions.

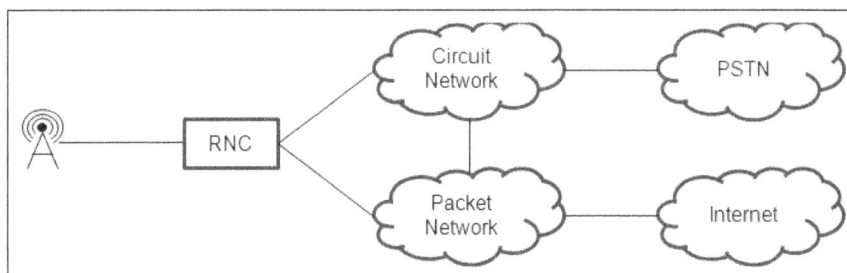

Structure of 3G Network.

The Radio Network Controller (RNC) manages radio resources for radio access, and Serving GPRS Support Nodes (SGSN) is responsible for management and support services in the PN. Gateway GPRS Support Nodes (GGSN) perform IP allocation for UE and support communication between PN and the external Internet network. Between each device uses a different protocol and tunneling. Between RNC and SGSN is called "Iu-PS" section usually used an ATM protocol, and SGSN and GGSN is called "Gn" section use the GPRS Tunneling Protocol (GTP). GTP is an IP-based protocol for tunneling between GGSN and SGSN. GTP can be categorized as GTP-U for the packet data, GTP-C for signaling, GTP ` (prime) for billing.

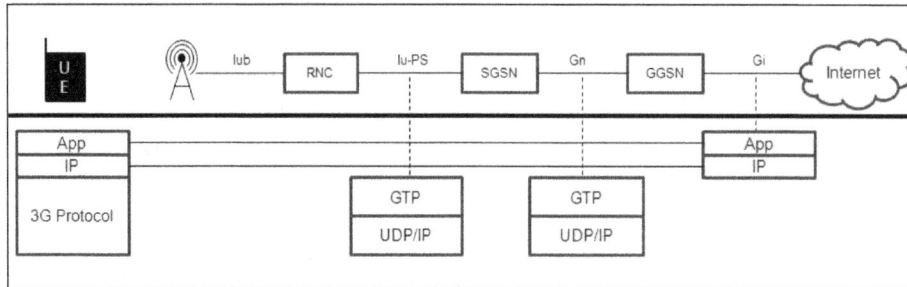

Structure of 3G Protocol Stack.

Mobile Terminal

Mobile terminals (aka UE; User Equipment) are information device that can communicate using 3G mobile networks. UE is representative of the smartphones but includes a variety of devices like laptop and tablet recently. In particular, variety UE that have not function for 3G mobile communication but have function for Wi-Fi communication can communication via 3G mobile networks using a tethering feature of smartphone.

Depending on the type of UE differed for the traffic. After all incoming traffic through 3G mobile networks are not only the smartphone but also multiple devices such as notebooks and netbooks. Thus, previously not seen in a mobile environment the traffic will now be observed in various forms.

Security Threat in 3G Network

Mobile networks have several features compared to traditional wired environment, it have relatively narrow bandwidth and limited radio resources, and complex signaling protocol for resource management. So these traffic which is not a problem in wired networks but mobile networks, it can be a threat.

Resource Consume of UE

Portable mobile device power management is very important. All work performed by the device is leads to power consumption, so if there are many unnecessary process, it will be leads to the consumption of power. In these cases, there are two broad, the first continuous communication, and the second is the high cost of process periodically.

First, in the case of resource consumption due to ongoing communication caused by an abnormal service continued communication and continuous communication with malicious host by infected with malicious code. In particular, the UE that is infected with malicious code continuous scanning to inside the network, the internal various UEs resources can be consumed unnecessarily at the same time.

In the second case, the allocation and release of radio resources for a short cycle can resource consumption. In order to communicate via mobile networks, the base station must assign a radio resource to the UE, but each base station can allocate the limited radio resources. Therefore, the base station allocates radio resources to the UE, after a period of time if there is no communication from the UE then releases the previously allocated radio resource.

The allocation and release of radio resource is very high task from the perspective of the UE. The state transitions are sketched in figure. The timer automatically turn off the wireless resource is called t, (t that can decide the value of the carrier), this t, if the attacker can constantly send packets over a slightly longer period then t, these can cause the radio resource re-allocation when released repeatedly. Once this process is repeated, the UE's resources are largely consumed.

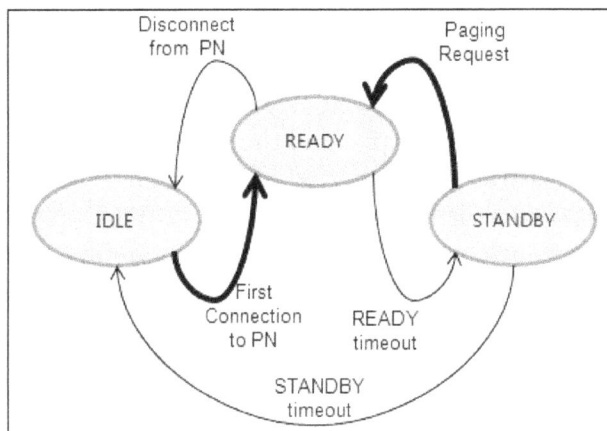

UE State on 3G mobile network.

Resource Consume of Mobile Network

Compared with wired networks, mobile networks have a complex protocol stack, so mobile network is higher cost required in order to processing than wired network. Recently, Mobile networks receptive traffic to the inside that existed in the wired network, such as unnecessary traffic and malicious traffic scanning. In a wired network, a specific target for attack is important, but the existence of such traffic in mobile networks, can be viewed as an attack on network.

Radio resources of a mobile network can be seen as an important resource for each base station has limited radio resources. Thus, one base station is limited to accommodate the UE, so UE need to assign radio resources when the communication only and if there is a certain amount of time with no communication and then return to make it available to the other UE.

But if attacker sent packet to any mobile device with anomaly timer that has a shorter period then normal timer t, the RNC keep radio resource to mobile device, and the UE will occupy the wireless resource and prevent use for other UEs. A small number of these abnormal communications from the UEs may not be serious, but voice and data services may fail of the area that covered by single base station if a large number of UEs rise at single base station.

The Abnormal Termination of Connected Service

The Messenger service is the most common connection-oriented services. In order to use instant messaging services connect with the server must be maintained. This connection to the regular communication is maintained for a period of time if you do not have a connection attempts to reconnect. The message service in a mobile environment while having the advantage of mobility has become a major mobile service. Communication between the PC as the existing constraints of mobility, but the mobile environment by using the 3G data network, while having high mobility can send and receive messages in real time.

Mobile messaging services emerged as the leading mobile killer service became mandatory. However, the mobile network takes a serious problem by the abnormal termination of service if messaging service is designed without consideration for the mobile environment. Abnormal termination of service, the UE will cause a continuous retry. In the attempt to reconnect the individual UE position is not a big problem, but the sides of mobile network it will be centralized at the request from the whole UE. And after a certain time, a time period of attempts to reconnect server is getting a short cycle and eventually all UE are constantly try to reconnect at the same time.

In this process, potential threats can be viewed as a three-part. In the first, UE resource is consumed by the repeated reconnection attempts, and the second occupying the wireless resource by repetitive reconnect to the service as short a period of time. The third, the RNC and the SGSN will be a failure by massive retries in a short time. In particular, centralized massive retries a short time can make a DDoS attack effect such as "SYN flood" attack to mobile network. In particular, a RNC is closely related both a data service and voice service, so the failure of RNC will lead to the failure of both data and voice services.

Counter Measures

Compared with wired networks, wireless networks have a variety of security threats because a special feature of mobile network such as limited radio resource and narrow bandwidth. Although, one could apply of existing security equipment which IP based, but it cannot cover the core network in 3G mobile network. Therefore, we need optimal security system for the 3G mobile network. In figure, the security system can be divided into three parts, the first traffic acquisition system, the second abnormal traffic detection system, and finally control system that can monitoring and control for detected mobile device based on detection information.

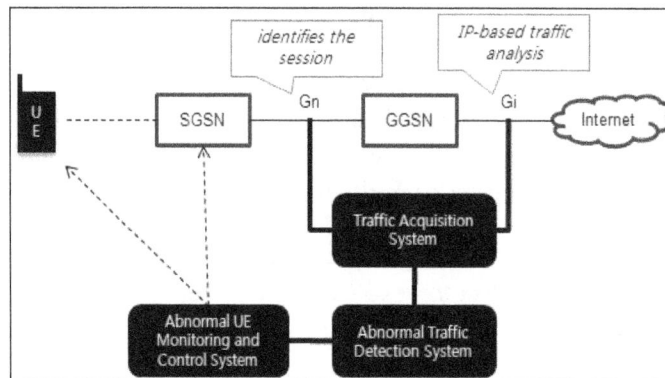

Framework for optimal security system for the 3G mobile network.

Traffic Acquisition System

The role of traffic-acquisition system (TAS) is to collect useful information from the traffic in PN. The Acquisition point will be Iu-PS, and Gn, Gi which main interface of PN and TAS designed able to analysis each communication session.

The main protocol architecture include: The main protocol are GTP-C, GTP-U, GTP`, GTP-C protocol use control the states of UE, GTP-U protocol use practical send and receive data. GTP` protocol use create CDR is used for the billing data.

In order to analyze end-to-end communication must see the inside of the tunnel because tunneling between the components of mobile network. In addition, the header for tunneling causes reducing the size of the entire PDU and fragmentation of packets. This phenomenon makes packet reassembly difficult. Therefore, we consider identifies the session at Gn interface, and IP-based traffic analysis at Gi interface. If traffic acquire at other interfaces except Gn, Gi interface, these case are on the costly disadvantages because required more financial and technical cost while more detailed monitoring is possible about the signaling and data traffic.

Abnormal Traffic Detection System

Abnormal traffic detection system consists of the two engines. One engine detects the engine has already revealed attack and other engine detects unknown abnormal traffic.

In terms of the network anomaly detection, there are false-positive and false-negative on the premises, so we need the clearly detected for known attack and more fine-grained analysis for the unknown anomaly attack detection. Anomaly Detection studied previously is based on the analysis of one-dimensional distributions of certain features across individual mobile users. However, the rapid growth of the mobile environment, we need to be verified result for depending on the nature of the carriers of each country.

In addition, we need consider recent trends in mobile services, and research should proceed according to the type of service, it is basis of selective detection measures.

Abnormal UE Monitoring and Control System

The monitoring and control system is responsible for control phenomenon caused by abnormal UE, based on detection information. Monitoring of the UE due to issues of privacy, collection and detection, and control systems cannot identify who the actual users, so only need to uniquely identify users by other information such as hash table, and also blind payload of the packet.

We need to control a UE which serious affect to availability of mobile networks. The control method of UE are use the delete function of the Android`s official provides that "Remote Application Removal", or sinkhole the destination of malicious traffic as a basic countermeasures may be considered.

In addition, if an emergency situation, force blocking for data communication or control of HLR to temporarily prevent the authentication of the UE as an extreme way. But more important, control of the UE is a forced act for the actual user, so we need sensitive approach.

Security of Software Defined Mobile Networks

The evolution to 5G and future mobile telecommunication networks is characterized by a significant surge in demands in terms of performance, flexibility, portability, and energy efficiency across all network functions. Software Defined Mobile Network (SDMN) architecture integrates the principles of Software Defined Networking (SDN), Network Function Virtualization (NFV) and cloud computing to telecommunication networks. The SDMN architecture is designed to provide a suitable platform for novel network concepts that can meet the requirements of both evolving and future mobile networks.

The underlying principle of the SDN architecture is the decoupling of the network control and data planes. Using this principle, network control functions are logically centralized and the underlying network infrastructure is abstracted from the control functions. The introduction of NFV offers a new paradigm to design, deploy and manage networking services based on the decoupling of the network functions from proprietary hardware appliances, and providing such services on a software platform. However, the separation of control and data planes as well as the virtualization of network functions and programmability introduces a number of novel use cases and functions on the network. This will further usher in new stakeholders into the networking arena and hence will obviously alter the approach to security management in 5G and future telecommunication networks. Several proposals are available for securing general SDN networks and SDMNs. However, none of these solutions provide a unified solution to secure future 5G SDMN backhaul network. Therefore, it is needed to define comprehensive security architecture for 5G SDMN networks.

SDMN Architecture

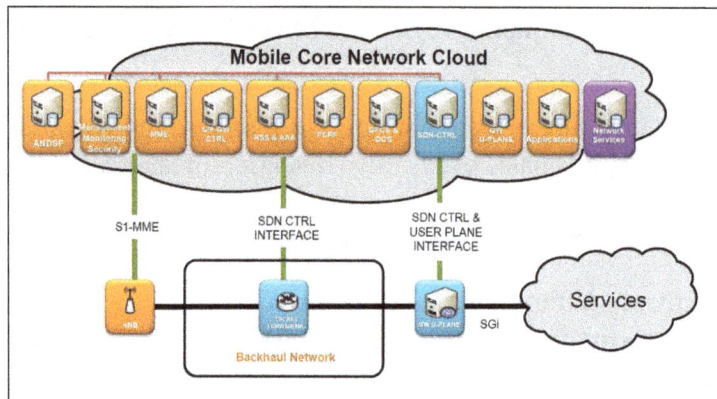

The consolidated SDMN architecture.

SDMN architecture integrates the core principles of SDN, cloud computing, and NFV into a design of programmable flow-centric mobile networks providing high flexibility. This modification is of significant improvement to the current LTE 3GPP (3rd Generation Partnership Project) networks. It offers benefits such as a uniform approach to Best Effort and Carrier Grade services, centralized control for functions that benefit from a network wide view, improvement in flexibility and more efficient segmentation. It also provides an enabling platform for automatic network management, granular network control, elastic resource scaling and cost savings for backhaul devices. With SDMN, resource provisioning is done on-demand, hence allowing elastic resource scaling across the network. With these attributes, SDMN becomes the latest innovation in the field of telecommunication. A consolidated illustration of the SDMN architecture is presented in figure.

In this architecture, traditional legacy control functions which include the MME (Mobility Management Entity), the HSS (Home Subscriber Server), the PCRF (Policy and Charging Rules Function) and the control planes of S/P-GW (Serving/Packet Gateway) are all run as SDN applications atop the mobile network cloud. With this approach, the user plane will consist of SDN enabled switches and devices placed in strategic locations on the network.

SDMN applies to both LTE and 5G network. Currently, 5G is planned to meet the needs of both the consumer markets and new massive machine-to-machine communications with tailored support for ultra-high reliability applications. Based on current 5G standardization activities, the assumption is that the 5G core network will be based on SDN. It is also planned that the core network will be sliced for better isolation and tailoring to the particular requirements of the market segments. The exact set of network functions in each slice can vary.

Security Threats in SDMN

As an ever growing share of Internet use is over mobile networks, inherent Internet threats such as ease of Denial of Service (DoS) attacks, source address spoofing and distribution of malware apply to mobile networks as well. Similarly, SDN and NFV have their own security limitations, and deploying these concepts in mobile networks without considering their inherent limitations will further elevate the security challenges. Hence, the separation of planes, aggregating the control functionality to a centralized system and running the control functions in the cloud as in SDN will open new security challenges for SDMNs. For instance, the communication channels between the isolated planes can be targeted to masquerade one plane for attacking the other. The control plane is more vulnerable to security attacks, especially to DoS and DDoS (Distributed DoS) attacks, because of its centralized nature and global visibility and can become a single point of failure. Since the networking paradigm of future mobile networks is converging towards software-based networking, operational malfunctioning or malicious software can compromise the whole network by getting access to the control plane. Some of the known security challenges in SDMN are summarized in table.

Table: Summery of security threats in SDMN architecture.

SDMN Layer	Types of Threat	Threat Reason and Description
Application	Lack of authentication and authorization.	Possible huge number of (third-party) apps.
	Fraudulent rules insertion.	Malicious applications generated false flow rules
	Access control and accountability.	Lack of binding mechanisms for apps.
Control	DoS, DDoS attack, Controller hijacking or compromise.	Visible nature of Ctrl-plane.
	Unauthorized controller access.	No compelling mechanisms for enforcing access ctrl on backhaul devices.
	Privacy of communications.	Attacker with access to controller can command to fork any flow at any point to a VNF function anywhere where it can analyze the content breaking confidentiality of communications.
	Scalability or availability.	Centralized intelligence

Data	Fraudulent flow rules.	Lack of intelligence
	Flooding attacks.	Limited capacity of flow tables.
	Controller and DP switch masquerading.	Lack of strong authentication
Ctrl-Data Interface	TCP-Level attacks.	TLS is susceptible to TCP level attacks.
	Man-in-the middle attacks.	Optional use of TLS and complexity in configuration of TLS
App-Ctrl Interface	Illegal controller access, policy manipulation and fraudulent rule insertion.	Limited secure APIs, lack of binding mechanisms b/w Apps and controller.

Since, SDN is considered to enable innovation in communication networks, bring flexibility and simplify network management, research efforts are going on for the deployment of its concepts in mobile networks. From security perspectives, SDN will enhance network security for two main reasons. First, it centralizes the network control plane that will provide global visibility of the network state and traffic behavior. Second, SDN brings programmability into communication networks through programmable APIs in the data forwarding elements. These two aspects enable SDNs to facilitate runtime network monitoring with quick threat identification, faster response systems, easy security policy alteration, and fast security service insertion without individual device configurations. Therefore, several security systems development proposals for SDN-based networks are proposed such as FRESCO, FSL and splendid isolation. These mechanisms can be used to develop mobile network specific security techniques. Various approaches are also suggested to secure SDNs due to its inherent limitations. These technologies and proposals are listed in Table which presents those security solutions with the type of security and the target SDN plans and interfaces.

Table: Proposed security mechanism for general SDN networks.

Security Type	SDN Layer/ Interface
Threat detection and mitigation.	Ctrl, App- Ctrl
App debugging, flow rules inspection.	App, Data
Flow rules verification, Configuration verification.	Ctrl, Data
Flow policy verification, catch bugs in OF programs.	App, Ctrl
App testing and debugging.	App
Conflict resolution, authorization, security audit system.	Ctrl, Data, App- Ctrl
DDoS detection, Controller resilience.	Ctrl, Data
Link monitoring.	Data, Ctrl, Data
Find contradictions in flow rules, authorize applications.	Ctrl, Data
Controller availability, network monitoring.	Ctrl, Ctrl, Data
Access control and dynamic policy enforcement.	Ctrl, Data

There are several proposals for securing SDN based mobile network architectures from a particular security threat. For example, proposes vulnerability assessment methodology for SDN based 5G network architectures. Similarly, proposes leveraging SDN to strengthen authentications security and protect privacy during handovers in 5G networks. The proposed mechanisms in also simplifies the handover authentication in heterogeneous 5G networks leveraging on global visibility attained by the centralized control platform of SDN. However, there is no unified solution to the future 5G mobile networks that provides security to the whole backhaul and the core networks along with the transport channels.

The SIGMONA project proposed SDN-based mobile network architecture. Then telecom-specific security requirements which gathered for a consolidated security architecture that efficiently secures the whole SDN-based mobile network that we call SDMN. This paper presents the SDMN backhaul security architecture proposed in SIGMONA with its validation results.

Security Requirements for SDMN

In addition to challenges from new technologies: SDN and NFV, the growing popularity of smartphones, rising mobile broadband volume and sophistication of malware exposes the mobile-terminals and their networks to the attacks of the fixed networks, such as source address spoofing, unwanted flows, malicious traffic and DDoS. However, compared to their fixed counterparts, i.e. laptops and desktops etc., mobile terminals are constrained by computing resources, storage and battery lifetime. This is even truer for some of the new devices envisioned to connect under 5G, such as sensors etc. which could be even resource constrained. This deters deploying the host-based security solutions on the wireless hosts. Moreover, the host-only security would leave the backhaul network and radio interface unprotected against hacks, malicious flows and unwanted traffic from the Internet, taking a significant toll on network performance.

Taking a fresh look at the end-to-end principle, we state that a function that is not feasible in the end-hosts shall be left to the network. The new technologies and planned enhancements in the core network can significantly contribute to the security of the network as well as its users. For example, relying on the principles of SDN and SDMN can leverage the global visibility of the SDN controller on the underlying network to: a) enforce consistent security policies across the network; b) fine-grain handling of individual user flows and new flows in network; and (c) to dynamically react to evolving threats by forwarding updated firewall rules to the data-plane nodes.

To address these Internet threats, we argue that future mobile networks should:

- Limit the flow acceptance to verifiable sources, to tackle the problem of unwanted traffic, source address spoofing, and thus prevent resource exhaustion.

- Eliminate source address spoofing to attribute the evidence of misbehavior to the sender.

- Make it possible to aggregate misbehavior evidences under a stable source identity, and contribute towards using reputation mechanisms for improving the security of communicating entities.

- Under network stress, grant resources based on source reputation.

- Allow defining dynamic (reachability) policies for hosts, applications and services. The management and control of the policies will be in the cloud while enforcement takes place in standard data-plane nodes on trust boundaries. This is in contrast to the current mobile networks where policies are tightly coupled to physical resources and are not scalable to services/applications.

- Leverage the logically centralized controller to overview, analyse and manage the policy configuration of data-plane elements, in order to deploy a robust and consistent security policy across the network.

In addition, the deployment of existing and new mechanisms to SDMN requires that they are implemented and testbed for their compliance to SDN principles, because existing solutions could be difficult to deploy, manage and scale to secure SDMNs. We argue that the security solutions should:

- Optimize the network resource utilization for security functions.

- Leverage the existing research/work in network security to harden SDMNs against classical Internet attacks.

- Limit all the changes to network edges, and not require any mandatory changes in the end-hosts or protocols, to minimize the deployment challenge.

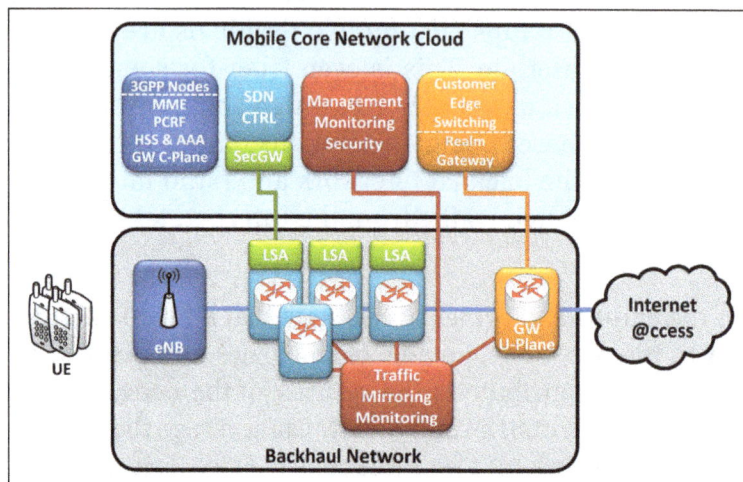

The proposed security architecture for SDMN.

Proposed Security Architecture

Given that most of the requirement specific to telecom architectures is tightly coupled with the control and data planes than with the application plane, hence, the proposed security architecture is geared towards securing the control plane, data plane and the Ctrl-Data interface (southbound interface). Figure presents the proposed security architecture for SDMN networks.

The proposed SDMN security architecture is a multitier security approach with five components, namely:

- Secure Communication (SC) Component.

- Policy Based Communication (PBC) Component.

- Security Information and Event Management (SIEM) Component.

- Security Defined Monitoring (SDM) Component.

- Deep Packet Inspection (DPI) Component.

Secure Communication (SC) Component

The SDMN architecture comprises of two main communication channels, the data and control channels. The data channel handles the transportation of the user communication data while the control channel handles the movement of essential control and signaling data between the data and control planes.

The major security concerns in SDMN communication channels are the lack of IP-level security and weak authentication between backhaul devices as shown in table. Existing SDMN communication channels are heavily reliant on higher layer security mechanisms like TLS (Transport Layer Security) /SSL (Secure Sockets Layer). A typical example is the widely used OpenFlow protocol which runs over a TLS/SSL based control channel. However, such higher layer security mechanisms offer no protection to information at IP levels. This leaves the communication sessions vulnerable to IP based attacks such as TCP SYN DoS, TCP reset attacks and IP spoofing. In addition, the TLS/SSL authentication mechanism is also exposed to IP spoofing and Compression Ratio Info-leak Made Easy (CRIME) attacks. These vulnerabilities buttress the need for secure communication mechanisms in SDMN architecture so as to mitigate against such threats.

To secure the communication channel of the SDMN architecture, we propose a HIP (Host Identity Protocol) based secure IPsec tunnelling architecture, this architecture helps to establish secure HIP tunnels between the controller and the DP (Data Plane) switches. The latest IP based telecommunication network (i.e. LTE/LTE-A) operators are heavily relying on IPsec tunneling and security gateway mechanisms to protect their backhaul traffic. Several versions of IPSec key exchange mechanisms are available such as Internet Key Exchange version 1 (IKEv1), Internet Key Exchange version 2 (IKEv2), IKEv2 Mobility and Multihoming Protocol (MOBIKE), Host Identity Protocol (HIP). However, it is not possible to implement these legacy IPsec tunneling mechanisms in SDMNs due to several limitations, such as distributed tunnel establishment, lack of centralized controlling, Point-to-Point (P2P) tunnel establishment, per tunnel encryption key negotiation, limited security plane scalability, lack of visibility, lack of access control and static tunnel establishment. The propose security mechanisms overcame these identified limitations.

Figure illustrates the secure HIP tunnel establishment under the proposed SC component.

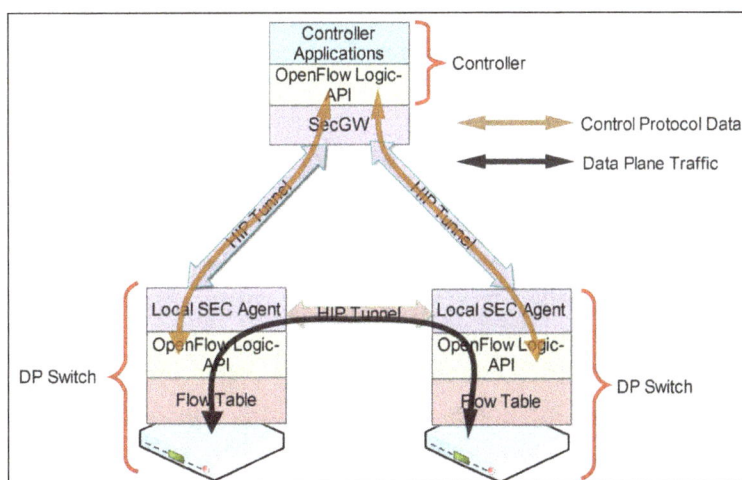

Secure communication channel.

Here we propose three key modifications to existing SDMN architecture. First, we introduce distributed Security Gateways (SecGWs) for securing the controller from outside network and mitigating against the odd of a single point of failure. SecGW is the intermediate gateway device between the controller and the data plane switches. It not only hides the controller but also reduces the security work load on the controller. Second, we added a new Security Entity (SecE) to control SecGWs and other security functions in SDMN. Third, we installed local Security Agent (LSA) application in each data plane switch to manage security related functions in the switch. The proposed solution is a bump-in-thewire mechanism and it does not affect the underling control protocols like Openflow nor other user-plane communication protocols. In Table, we compare various features of the proposed SC component with other security solutions.

TABLE: Comparison of proposed architecture with existing security mechanisms.

Property	TLS/SSL	IPSec with 1KEv2	IPSec with HIP	Proposed Architecture
Vulnerability of mutual authentication mechanism.	Medium	Medium	Low	Low
DoS attack prevention.	No	No	Yes	Yes
Support for seamless mobility of backhaul nodes.	No	No	Yes	Yes
Multihomed Support.	No	No	Yes	Yes
Centralized Controlling.	No	No	No	Yes
Point-to-Multipoint/Multipoint-to-Multipoint.	No	No	No	Yes
Visibility of traffic transportation.	No	No	No	Yes
Access Control.	No	No	No	Yes
Collaboration with other control entities.	No	No	No	Yes

Policy Based Communication (PBC) Component

The best-effort paradigm of the current Internet allows a host to initiate flows towards any destination address. Hackers often abuse this paradigm to launch attacks to their victims. Hiding under a spoofed address, hackers often bypass security mechanisms and also prevent tracing the attack back to its originator. The best-effort principle attempts its best to deliver the packets of the sender to the destination. However, because interests of the receiver do not always align with the sender, the destination receives unwanted traffic. Since SDMN proposes an all-IP based open network architecture, it is also vulnerable to unwanted traffic, DoS and source address spoofing similar to other IP networks such as Internet.

Cooperation is a proven mechanism to effectively curb the antisocial behavior in a population. We propose a two-tier cooperative approach to improve SDMN security and limit the extent of damage from Internet malpractices. The goal is to: 1) mitigate traditional attacks on SDMNs, i.e. DoS and source address spoofing; 2) encourage cooperation of all benevolent entities against the malicious sources; and 3) tracing as well as containing all the resources used by the hacker in attacks. First tier is achieved by establishing the required level of edge-to-edge trust using Customer Edge Switching (CES). The second tier involves the ubiquitous collection and attribution of the attack evidences within a trust domain. CES nodes will then use the consolidated evidences to black and grey list remote entities.

CES allows policy-based communication to mobile hosts. A CES node in principle replaces NAT at the network edge, and extends the classical stateful firewall into cooperative firewall. CES acts as a secure connection broker for hosts located in its network, and in contrast to the classical Internet matches the interest of the sender with the receiver prior to flow admission, or before forwarding a new flow on the Internet. The interests of the end-hosts are expressed as a policy, which could require stable source identity, verification of the sender credentials, i.e. via reverse DNS (Domain Name Server) lookup to more complex certificate validation, as well as the possibility of utilizing the private-transit links, not exposed to the public Internet. The negotiation of interests between CES firewalls for respective hosts effectively extends the classical statefull firewall functionality into cooperative firewall.

For mobile networks, CES offers many advantages: (a) end users will benefit from a network firewall in the cloud, instead of relying only on host-based security solutions on the mobile device for blocking unwanted traffic and common attacks. This (b) saves computing resources of the device; and (c) contributes to battery lifetime of wireless device, by preventing unwanted traffic from reaching to device and disturbing its sleep cycle; besides preventing d) cluttering of air interface and network. CES does not require changes in end protocols, applications, or any explicit signaling from hosts to maintain their network connection: NAT bindings, or connection states. The policy-based communication facilitated by CES means that all flows to the mobile hosts are admitted based on policy. Prior to admitting a flow, the outbound CES (oCES) node and the inbound CES (iCES) node will negotiate policies of the respective hosts via Customer Edge Traversal Protocol (CETP). In case of a policy matching, the connection state is inserted into data-plane to admit the user connection. Figure shows the deployment of proposed PBC component, (i.e. CES) which runs as an SDN application on top of the SDN controller.

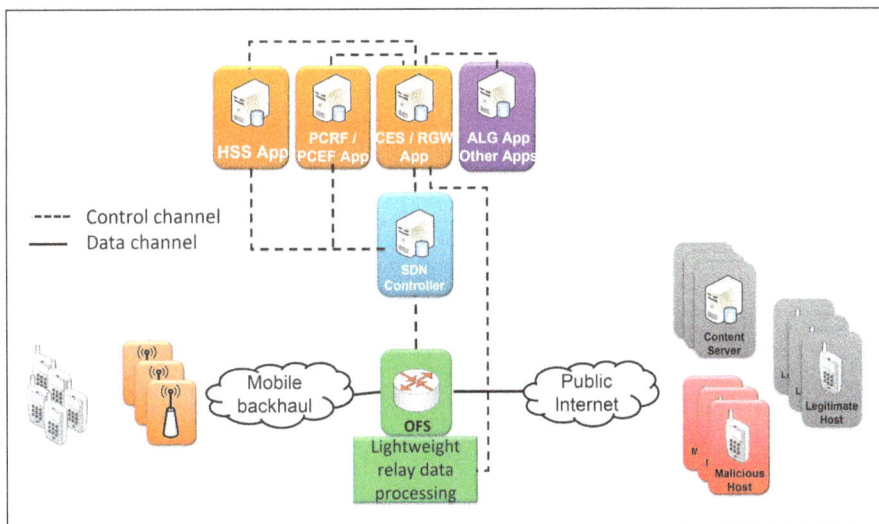

SDN oriented customer edge switching.

CES retrieves the necessary user policies and certificates from mobile network components, such as HSS or PCRF on demand. Upon successfully negotiating policy, CES interacts with the data-plane to insert the negotiated flows into the OpenFlow switches, and hence allows the end-to-end user communication.

CES provides backwards compatibility with legacy networks using Realm Gateway (RGW). For outbound connections, the RGW function is similar to NAT, however it admits inbound connection following a domain name query from the Internet towards a private host, leading to creation of a NAT binding and granting connectivity. We implemented a number of heuristic mechanisms to secure the interaction of the legacy Internet with CES/RGW. For DNS, these include classifying DNS servers into white, grey and black lists based on: service-level agreements, influx of DNS floods and resource assigning model. Whitelisting can be based on spoofing-free DNS/TCP channel, SLAs and use of private links between networks. The address allocation in the realm gateway is controlled by policy to limit or deny the resource allocation to any host, server or a given network. Upon a new valid host name query, the address allocation algorithm creates a half-open state with respect to sender which is elaborated into a full inbound NAT binding upon the first inbound SYN from sender. In the full binding state, the RGW applies the address and port dependent filtering on incoming packets relative to the client. We also define Service-FQDN to address the services running on end-hosts; this not only contributes to security but also increases the scalability of RGW. The naming scheme allows a stricter NAT binding by virtue of adding a port number to the half-open state, making it more resistant to attacks and improving the reuse of the (pool of) inbound public addresses.

RGW employs the TCP-Splicing mechanism to eliminate spoofing in the user connections admitted to the network. For this it leverages the SYN cookie algorithm to postpone the allocation of TCP resources until the sender is determined as non-spoofed. Consequently, the network is protected against spoofed flows, and resources are assigned to valid sources. We also leverage the SYN cookie algorithm to implement a bot-detection method that detects and mitigates SYN floods from non-spoofed malicious sources, repeatedly targeting the temporary half-open connection states or bindings in RGW.

The proposed PBC component offers several benefits to its adoption. Since it uses standard DNS requests as communication trigger and it can be deployed transparently to end hosts, i.e. as it requires no changes in end-hosts, protocols or applications. It offers a light-weight, host independent NAT (Network Address Translation) traversal solution to admit the inbound connections. The centralized operations of SDN can contribute to more fine-grained and informed decision making of the heuristic algorithms, contributing to a more secure environment in SDMN.

Security Information and Event Management (SIEM) Component

Network monitoring solutions come in different variants depending on what they measure and how they collect the data:

- Active Probing: Service-centric approach that collects data based on synthetic measurements such as ICMP Echo Requests, HTTP GET requests or specially crafted packets. Often these measurements attempt to measure properties of the network that would be impossible to capture from pure passive measurements and are arguably the only way to measure service availability.

- Device Polling: Device-centric approach that queries devices typically using SNMP (Simple Network Management Protocol), collecting interface status information, traffic volumes, device load, CPU, etc.

- Flow Collection: Solutions that collect traffic information from network devices such as routers/switches; traffic is aggregated in flows using e.g. Cisco Netflow and stored in disk for post-analysis. Flow data is easier to analyze and process than packet data, but provides less granular information.

- Packet Analysis: Usually involves a SPAN port from a switch or a network tap and extracts information from individual packets, including information from payloads through DPI.

- Log Analysis: Solutions that collect machine generated data typically in the form of log files (e.g. syslog) and present a query interface to correlate events across different types of systems, e.g. routers, web servers, load balancers.

Security Information and Event Management (SIEM) component collects flow information and takes the security ensuring actions such as access control list update, firewall update, flow table modification; rate bound enforcement and so on. SIEM provides Security Information Management (SIM) on the one hand and Security Event Management (SEM) on the other hand. SIEM component also collects event data from network infrastructures, applications and security devices. Although the SIEM component uses log data as the primary data source, it can also generate other forms of data such as NetFlow and packet capture. Data from such events are combined with other contextual information regarding the users, assets, threats as well as other perceived vulnerabilities. To ensure that data and events are correlated, contextual information from disparate sources are normalized and analyzed for specific purposes, be it monitoring of user activities, monitoring of network security events or compliance reporting. The SIEM component performs real-time security monitoring and historical analysis. It also provides support for investigating incidents and providing reports on performance. It also contains a security monitoring and event management that perform an analysis of security event data in real time focused on network events, and present security information in a consolidated Graphical User Interface (GUI).

Figure shows a high level overview of the SIEM component.

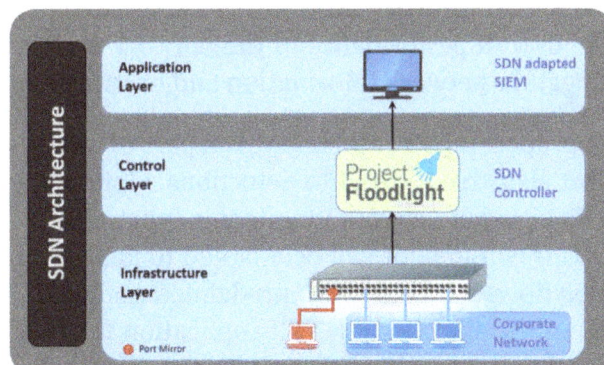

Security information and event management (SIEM) component.

It mainly consist two components:

- Security Sensor (Port Mirror): Responsible for gathering security information (i.e. through IDS (Intruder Detection System) features for security event detection) and reporting to the security server.

- Security Server (SDN Adapted SIEM): Responsible for collecting security information coming from deployed security sensors. The collected information is correlated and validated against predefined security policies for a final decision making. It optionally allows an automatic reaction to detected security events.

Therefore, proposed SIEM component offers the following features:

- Security Management features:

  ○ Security policies definition.

○ Countermeasures definition.

- Security Monitoring features:

    ○ Asset inventory – Availability monitoring.

    ○ Network monitoring (usage and latency).

    ○ Vulnerability discovery.

    ○ Event detection (intrusion, anomalies, etc).

Deep Packet Inspection (DPI) Component

SDMN enhances security by making it easier to implement counter-measures and isolate network parts when security problems are detected. On the other hand, additional software, components and interfaces required in SDMN open new opportunities for attacks by malicious agents. Security needs to be addressed on the network side as well as the mobile device side. Deep Packed Inspection (DPI) as part of Network Intrusion Detection Systems (NIDS) strengthens network security by detecting and tackling harmful traffic flows.

In SDMN, it is crucial that both applications and associated control elements are constantly aware of the conditions of the underlying infrastructure so as to guarantee optimal security at different levels. This is central to the overall performance of the network and can well be handled by the DPI. The DPI can routinely gather network information and channel it back to the control layer.

As illustrated in figure, the proposed DPI component is a part of an active monitoring probe that detects security threads. It is also able to react to detections. Optionally, the DPI component can work with mirrored traffic as a part of network monitoring functions. In both cases, the DPI component can be virtualized. This component will be adapted to analyze and detect diverse security threats related to application flows matching with predefined malware rule databases. The developed solution concentrates on the analysis of HTTP application flows. The detections are written to the local database and optionally provided directly to the other network functions. Currently, the proposed DPI component is not compatible with HTTPS flows. However, we assumed HTTPs traffics are protected at the application plane.

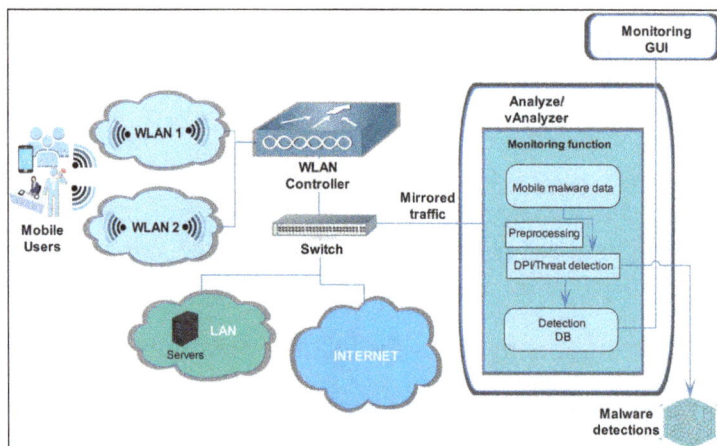

Deep packet inspection (DPI) component.

Security Management And Monitoring (SMM) Component

SDM component is designed to perform monitoring functions in SDN/NFV-based 5G mobile network architectures. It is able to monitor both virtualized and physical network environments in an economical and efficient way. Initially, the proposed SDM architecture is used only to monitor SDMN backhual networks. However, the proposed SDM architecture extend the capabilities of current SDN/OpenFlow features to provide the required level of monitoring capabilities in 5G backhual networks. Figure illustrates how SDM component is implemented in the 5G SDMN architecture.

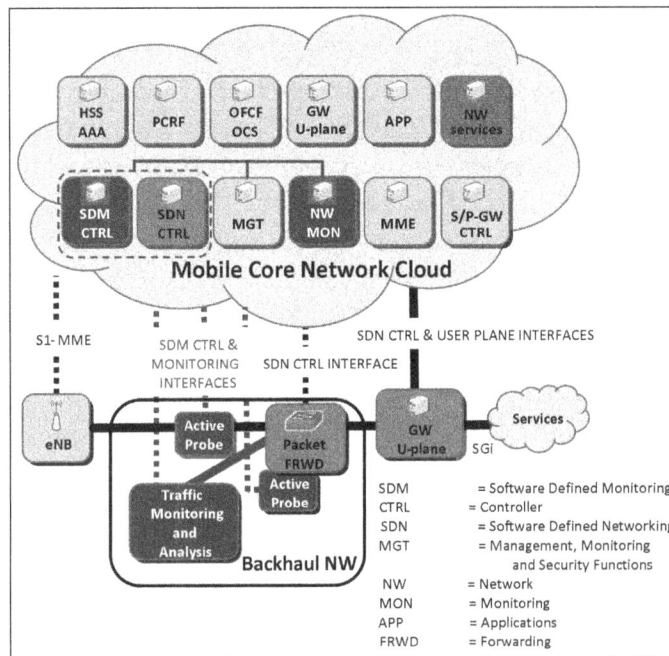

Security management and monitoring component.

The following modules and interfaces have been added to the SMM components.

- Modules:
  - Security sensor: An active monitoring probe used to detect information related to security and anomalous behaviors on the network. It also tries to mitigate detected attacks through the use of mechanisms such as filtering. Information collected by the security sensor may include general security properties and attack reports. Security sensor can be installed on the network elements or in network taps (passive network observation points).

  - SDM CTRL: An extension of SDN CTRL which allows the control of monitoring functions such as management of network monitoring appliances, traffic mirroring, traffic load balancing and aggregation. This module also attends to requests from network functions and applications. SDM CTRLs are usually distributed following either a peer-toper or hierarchical model. They inter-operate with the management/monitoring/security function and act as distributed analysis or decision points for the defined security policies (security SLAs).

  - Network monitoring: A virtual monitoring module which extends part of the traffic analysis to the cloud.

- ○ Traffic mirroring and analysis: A passive traffic monitoring device located at the backhaul to monitor variety of network functions.

- Interface:

  - ○ SDM CTRL Interface: Controls the use of monitoring resources, recuperating traffic or metadata for analysis. This interface allows monitoring requests to be sent to ascertain the status of the network, hence enabling applications and network functions to send requests for monitoring-based information, and monitoring functions can send status and recommendations.

The SMM component introduces a dedicated Software Defined Monitoring controller (SDM CTRL) to orchestrate the monitoring activities related to security that are performed by the security sensors (i.e., probes) deployed in the network and in the cloud. The SDM and SDN controllers can be separate modules or integrated into one module. The SDN CTRL also interacts with the routers implementing the SDN CTRL interface to manage the traffic (e.g. redirect traffic to the security appliances) and recuperate certain information. The SDM controller interacts with the security appliances or probes implementing the SDM CTRL interface to manage them and recuperate the metadata part of the traffic or verdicts. This information can be used by the network monitoring function in the cloud to perform analysis and trigger mitigation actions; of by the other network functions/services and applications.

Security sensors used in SMM component can be passive (not disrupting traffic) or active (in the data path to perform online countermeasures); and can either be installed in existing network elements or in dedicated security appliances. The probes analyze network traffic, correlate information from different sources and produce meta-data and verdicts that can then be used by a centralized decision point and by the different network functions.

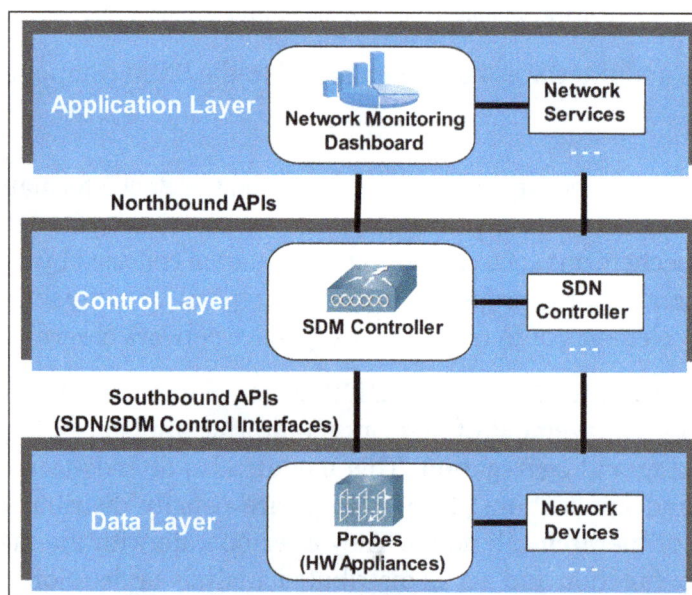

Security management and monitoring (SMM) component in three layer SDN architecture.

Figure shows how the components of the proposed architecture map to the three-layer mobile SDN architecture proposed by Open Network Foundation (ONF).

Testbed Implementations and Results

Here, we implement a proof-of-concept prototype on a testbed for the components of the proposed architecture in four sets of experiments. We then provide a performance evaluation for each component.

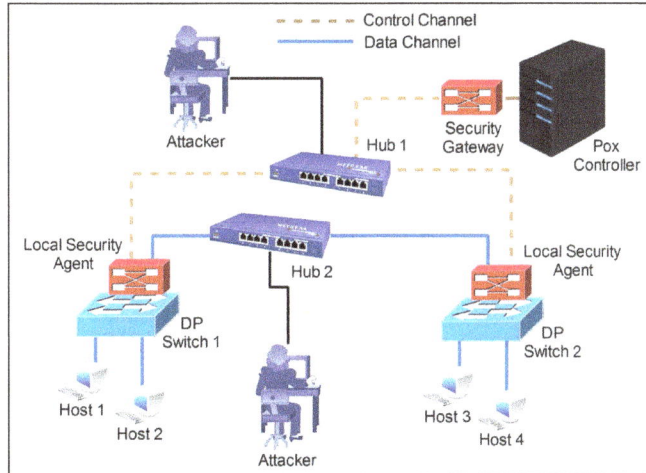

The layout of the experimental testbed for secure communication (SC) component.

The first set of experiments was for the secure communication component. In this experiment, we evaluated the performance penalty of this component in terms of throughput, jitter and latency. We further measured the capability of the proposed architecture to protect the communication channels against common IP based attacks like TCP SYN DoS and TCP reset attacks. We used OF protocol with TLS/SSL session as reference for the control channel. Figure illustrates the preliminary testbed components for this experiment.

As shown in this figure, the testbed contains two Data Plane(DP) switches, an SDN controller and two hubs. We used the latest version of POX controller as the SDN controller and OpenVswitch (OVS) version 1.10.0 virtual switches as DP switches. We have used four virtual hosts as users. For each OVS, two virtual hosts were connected. We used two D-LINK DSR-250N routers to connect the controller and the switches. For this experiments, we kept out-band control channel. We modeled the Security Gateway and LSAs using OpenHIP. We used IPERF network measurement tool to measure the performance in terms of throughput and latency. We finally connected an attacker to each hub for each scenario of the experiment; the attacker operates from an i5-3210M CPU of 2.5GHz processor laptop.

Table: The simulation settings for the IPERF.

Parameter	Value	Value
Protocol	UDP	TCP
Port	5004	5004
Buffer size	Default (1470 kB)	Default(1470 kB)
Packet size	Default (1470 B)	Default(1470 B)
TCP window size	–	21.0 KByte
Report interval	1 s	1 s

The experiment settings of IPERF testing tool is presented in Table.

TABLE: Data channel performance without attack (normal operation).

Performance Metric	Existing SDMN Data Channel	Proposed Secure Data Channel
TCP Throughput (Mbps)	93.5514	91.8054
UDP Throughput (Mbps)	95.2845	92.3828
Latency (ms)	36.6514	37.6452
Jitter (ms)	0.34522	0.4651

Table presents the performance of the data plane under each architecture. We ran each experiment for 500 seconds and recorded the average values of the outcome.

The experiment results presented in table indicated about 2% decrease in TCP and UDP throughputs for the proposed secure channel. In addition, we observed a 3% increase in latency when compared to existing SDMN data channel. This reduced performance of the network is caused by the extra layer of encryption added to the proposed secure channel. Notwithstanding, the addition of IPSec accelerators can help to further boost the performance of this architecture and minimize the deficiency caused by the extra layer of encryption. More recent Intel processors are capable of supporting such IPsec acceleration leveraging on external accelerators and/or using new AES (Advanced Encryption Standard) instruction sets.

For the next experiment, we added a TCP SYN DoS attacker to the data channel (Hub2). We ran each experiment for 500 seconds while launching attacks between 100 and 200 seconds time interval. Table shows the average performance of each architecture.

Table: Data channel performance under TCP DoS attack.

Performance Metric	Existing SDMN Data Channel	Proposed Secure Data Channel
TCP Throughput (Mbps)	72.1945	91.51564
UDP Throughput (Mbps)	74.4656	92.4551
Latency (ms)	548.14854	37.5146
Jitter (ms)	5.1495	0.4301

The outcome of the experiments recorded in table clearly shows the vulnerability of current SDMN channel to TCP DoS attacks. The experiment results show a 20% drop in throughput for both TCP and UDP in current SDMN data channel. The percentage drop in throughput is directly proportional to the percentage of time during which attacks were launched in reference to the overall experiment duration. We therefore conclude that current SDMN data channel is highly vulnerable to DoS attack, given that the effect of the attack on throughput lasted for the whole duration of the attack. Moreover, our experiment results also showed a 14 times increase in both latency and jitter using current SDMN data channel compared to normal operation. However, using the proposed secure channel, we experienced similar performance as in normal operation. Thus, we verify that the proposed secure channel is capable of securing the SDMN data channel from potential DoS attacks.

Table: Control channel performance under TCP DoS attack.

Performance Metric	Open Flow with TLS/SSL	Proposed Secure Control Channel
Connection Establishment Delay (ms)	58.3224	135.4165
Connection Establishment Delay under TCP SYN DoS Attack (ms)	-	135.9145
Flow Table Update Delay (ms)	30.85645	32.1573
Flow Table Update Delay under TCP Reset Attack (ms)	-	32.2472

In the next experiment, we orchestrated a TCP SYN DoS and Reset attack on the control channel (Hub1). We then recorded the connection delay and flow table update delay experienced between the controller and the DP switch 1. We ran each experiment 25 times and recorded the average performance of each architecture. Table shows the outcome of this experiment.

The experiment outcome presented in table shows a significant increase in connection establishment delay using the proposed secure channel. We observed an additional latency coming from the extra HIP tunnel establishment between LSA and SecGW. We also observed a 4% increase in flow table update delay using the proposed secure channel under a steady state of operation (i.e. after establishing the connection). This deficiency in performance comes from the extra layer of encryption when using the proposed secure channel.

The experiment results shown in table also shows how vulnerable the existing SDMN control channel is to TCP DoS and reset attacks. We observed that it was not possible to establish connections with the controller during TCP SYN DoS attacks and during the TCP reset attack it was not possible to update the flow tables. However, using the proposed secure channel, we observed consistent performance, hence resistant to those attacks. This confirms the ability of the proposed secure channel to secure the SDMN control channel from IP based attacks.

In the second set of experiments, we aim to measure the performance of the PBC component, as well as determine the strength and effectiveness of its security. The proposed PBC component consists of CES/RGW and is implemented in Python. The prototype is developed as CES proof-of-concept. Figure presents the implementation of our CES testbed, which is built in Linux environment and employs control/data plane split architecture. The testbed has two private networks that are respectively served by CES-A and CES-B. The edge of each network bears a data-path element, which enforces the rules generated by the control plane and forwards the new flows towards the CES function at the control plane, i.e. for connection admission, policy negotiation or security analysis. The nodes in the setup are implemented as linux containers and are connected using basic Linux networking support. Tests were conducted with a total of 8 simulated hosts.

Each network edge has two external interfaces: a) a public interface to receive traffic from the legacy Internet; and b) a more secure private-transit link to receive flows from hosts in a different CES network. The setup also contains few nodes in the legacy Internet to launch typical Internet attacks or abuses for testing the CES/RGW security. Under this setup, the PBC is testbed for both its use cases: 1) policy-based communication between CES nodes; and 2) inter-operability with legacy IP networks, using Realm Gateway (RGW) functions.

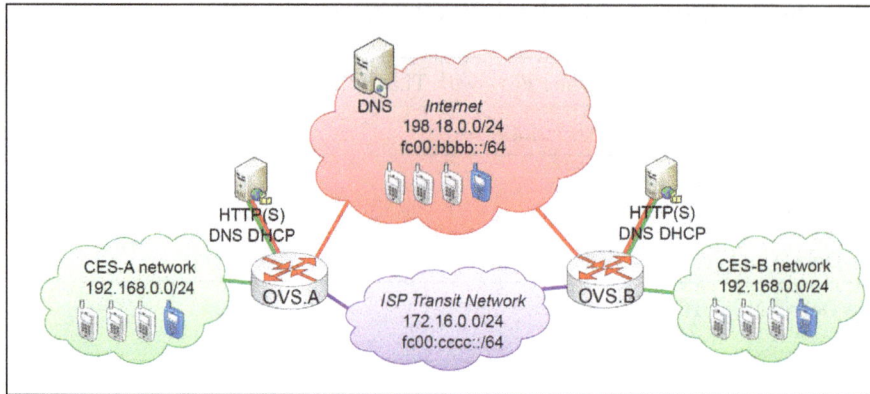

Layout of the experimental testbed for policy based communication (PBC) component.

Table: Security testing of CES policy negotiation.

Metrics	Testing Result	
Signaling round trips (RTTs).	1-RTT	2-RTT
Connection establishment delay (msec).	80	145
Proof-of-Work sender's delay (msec).	3	
Proof-of-Work receiver's delay (msec).	0.001	
Certificate/Signature computation (msec) 1st packet.	2	
Certificat/Signature verification (msec) 1st packet.	1.8	

Table presents results of CES security testing, in terms of the processing delay of the security mechanisms. The use of proof-of-work mechanism allows CES to push the burden of communication to the sender, such that sender invests more computing cycles than the receiver. This also effectively eliminates source address spoofing in the admitted flows, failing spoofed sources from leaking traffic into the private network.

The use of CES certificates (at CETP layer) coupled with signed CETP header is used to authenticate the remote node as CES. The mechanism leverages an object identifier in X.509 certificate to uniquely define CES certificates, and identify the remote node as CES (possessing the private-key) due to signed CETP headers. The mechanism is triggered upon the first flow from a new source and ascertains if the remote node is a valid CES. The testing revealed that only flows from valid CES nodes are admitted into the network. A CES node based on its policies decides whether to accept an inbound flow or request the sender for additional details, which may result in another round of policy exchange. The negotiation of policies completes in either one or two round trips and results in either: a) success; or b) failure depending on policies. The subsequent connections from the sender reutilize this validation result.

Having negotiated the CES policies, the subsequent flows from the sender only undergo one or two round trips of the host-to-host policy negotiation. A typical host-to-host user flow establishes after 80 msec or 145 msec delay incurred by 1-RTT or 2-RTTs of the host-policy negotiation, respectively. However, due to additional round of CES-policy exchange on the first inbound flow from the sender, the first host-to-host flow establishes in 220 msec for 1-RTT and 300 msec for 2-RTTs of the host-policy negotiation. Since we measured the connection setup on zero-latency links, one must add edge-to-edge latency of the real networks to get the actual connection setup delay. To account for network uncertainties, CES state machine can absorb any host retransmissions while the CETP process is still concluding.

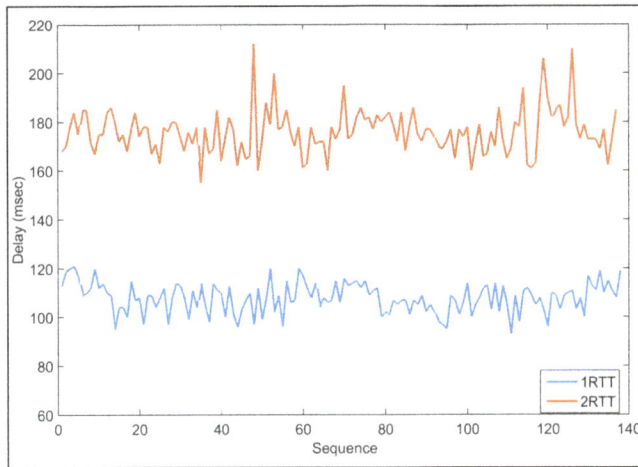

Delay induced by CETP policy negotiation on forwarding of
the first packet of the user-data connection.

Figure illustrates the connection setup delay of nearly 80 connections, using CETP host policies of varying complexity. The figure reveals that less complex policies are negotiated quicker than more complex policies that result in another round trip. Most of the presented delay in the figure is the due to slow control/data plane interaction, while the policy processing by CES is carried out in the order of milliseconds. In future, we aim to improve the CES-to-CES signaling by direct CES-to-CES control plane communication, and then synchronizing the negotiated user connection to the data-plane.

Figure shows the impact of resource allocation model on RGW on the event of a DNS flood. The model prevents the exhaustion of the address pool resources by rate limiting the DNS sources and by limiting the resources available to grey-listed DNS servers. By default, the servers that do not meet the SLA defined for trusted sources are grey listed. This results in higher availability of address pool resources to legitimate DNS servers and clients, particularly under load conditions. Our testing of RGW revealed that TCP-Splicing completely eliminated spoofing, and no spoofed source could leak traffic into the private network or claim a user connection. A future version of the prototype aims to employ SYN proxies instead of TCP-Splice, since they are optimized to handle millions of packets per second.

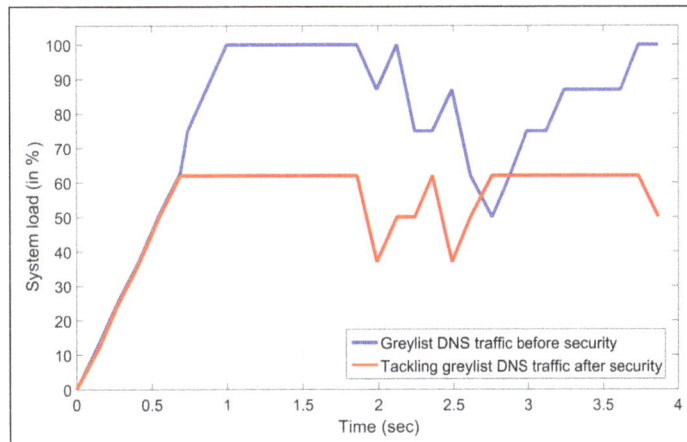

Tackling DNS flood from greylisted DNS servers

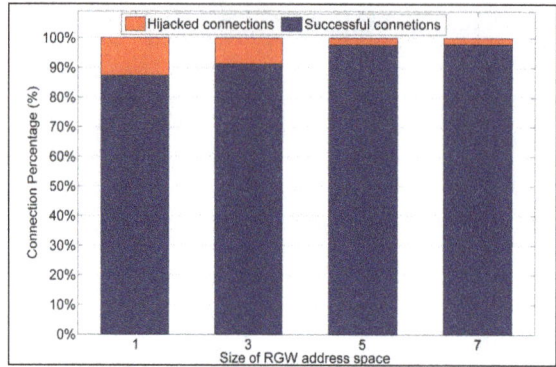

RGW security against non-spoofed floods.

Figure presents an evaluation of the bot-detection algorithm which aims to filter floods from non-spoofed sources. The figure shows the impact of increasing the pool of inbound public IP addresses on RGW security. The figure shows that the security offered by the bot-detection algorithm is more effective if the attack surface for the hackers is larger.

In the third set of experiments, we validate the performance of SIEM and SMM components. The experiment testbed is presented in Figure. We used Mininet v2.2.1 the network emulation environment and OpenvSwitch v2.3.1 for the deployment of SDN switches. Floodlight v1.1 was used as the SDN controller. Security monitoring and management elements such as SDN adapted SIEM, security sensor and security server was connected via a legacy switch. S1, S2, and S3 were virtual SDN switch which were implemented as OpenvSwitches. RO (Route Optimizer) deals with a virtualized element for routing purposes. The test network had been segmented into four LANs depending on the nature of their services. These segments have different security requirements. Tests were conducted with two users:

- DMZ LAN: It includes services exposed to Internet.
- Security LAN: It includes security services, such as the security sensor.
- Server LAN: It includes internal services.
- Client LAN: This is the end-user network.

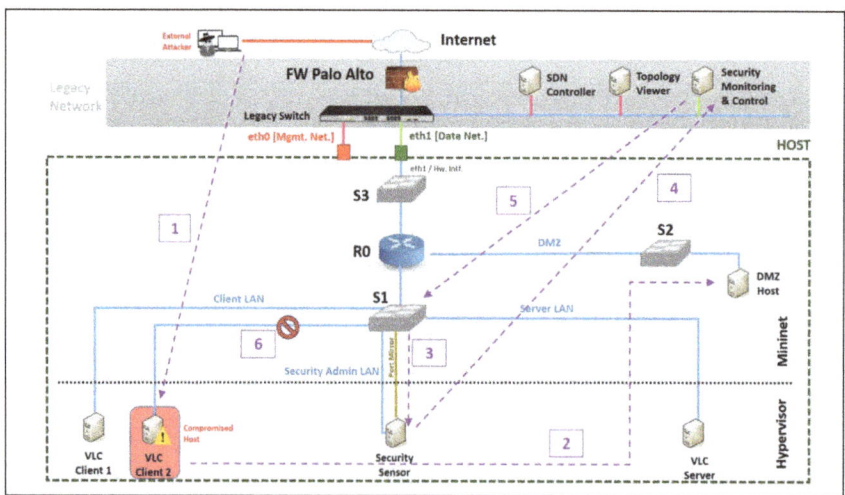

The layout of the experimental testbed for security information and event management (SIEM) and security management and monitoring (SMM) components.

In this testbed, a security use case had been defined as a proof of concept to show how SIEM and SMM components help detecting and isolating insecure network devices, before they can negatively affect the rest of the network. Upon discovering a potential threat by SMM, the SIEM identified the problem and automatically performs the previously considered or planned reactions to mitigate it, by interacting with the Northbound API of the SDN controller. After the threat had been resolved the SIEM software allows the affected devices to rejoin the network.

For the purpose of this use case, a VLC server streamed video in the server LAN and several VLC Clients were consuming this video from the Client LAN.

- The "VLC Client 2" has been compromised by an external attacker.

- The compromised host tries to extend the attack by launching a network discovery process over the internal networks, the DMZ LAN in this case.

- The suspicious traffic of the network discovery is detected by the security sensor that is sniffing all the traffic crossing the virtual switch S1.

- The Security sensor reports this security event to the Security Server, located in the legacy network.

- The Security Server processes this event matching against a predefined security policy that tells him to immediately block in S1 the connections to this host.

- Security Server processes the event and it is correlated in Security Server. The matching security policy (which is built on the server side) triggers an action that injects from Security Server to the SDN Controller via the NorthBound API. Then, the SDN Controller sends a flow table update message to the S1 to drop all the traffic related to the compromised host.

In the first case, SIEM performed cyber-attack detection by considering a unique source of information from the security sensor. The test consisted on detecting a port scan followed by sending ten echo requests (pings) that was detected by the security sensor running snort in a virtual machine deployment as described above. Different measurements have shown and compiled to an average to have a perception on the results of the different response times and latencies introduced. Figure represents the delay time, which was measured in seconds, between the attackers began to carry out the attack to the time that the attack had been blocked.

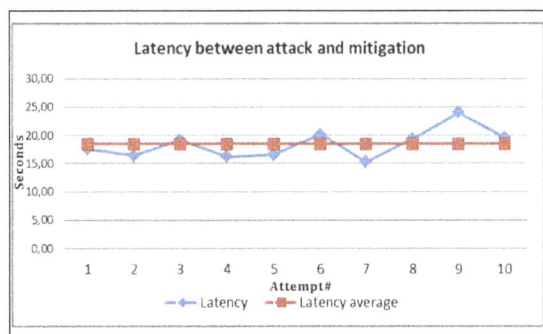

Latency between attack and mitigation. Case 1.

Figure represents the delay time between detection by the correlation engine generating the alert and the attacking device being blocked.

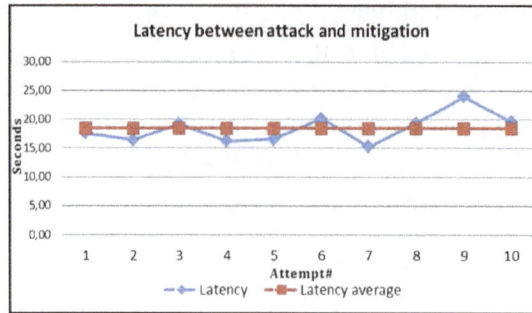

Latency between detection and mitigation. Case 1.

In the second case, SIEM considered different sources of information from the security sensor. In this case, the response times had been evaluated for a case of simulation of a botnet where a host took the control of another host in the network and proceeded to download a malware file. The compromised host and the attacker host generated a network connection and it was identified by the security server detecting an outgoing connection from the botnet web server. The compromised host downloaded a malware file that was also detected by the security sensor. In addition, a malware engine analyzed the downloaded file and assigned a score, in order to determine if it was considered malware.

Figure represents the delay time between the time the attacker began to carry out the attack to the time that the attack had been blocked.

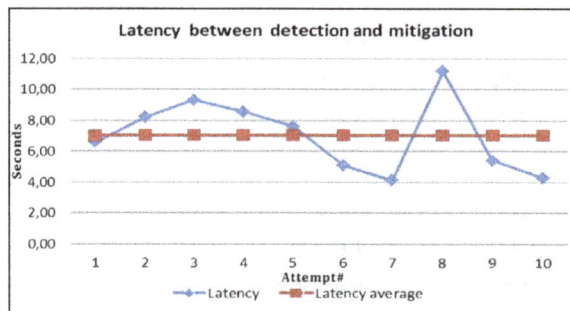

Latency between attack and mitigation. Case 2.

Figure below represents the delay time between detection by the correlation engine generating the alert and the attacking device being blocked.

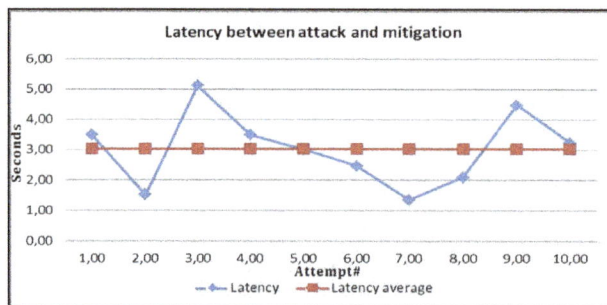

Latency between detection and mitigation. Case 2.

The results of the validation show that it is possible automate mitigation and reaction actions in SDMNs by providing countermeasures and mitigation actions directly using RESTful API in a SDN controller.

The result of the validation also evidences that multiples sources of information can be combined and help to provide more accurate and rapid detection on cyber-attacks scenarios demonstrated. Improving performance combining multiple sources will be crucial for future and further work.

In the fourth set of experiments, we validate the performance of DPI component. Figure illustrates main components of the developed monitoring prototype. The threat detection is based on malware fingerprints that are compared with monitored online traffic patterns as part of DPI analysis. Possible malware detections are then written to the local database with the other analysis data produced by the DPI component. Tests were conducted with a total of 10 simulated hosts.

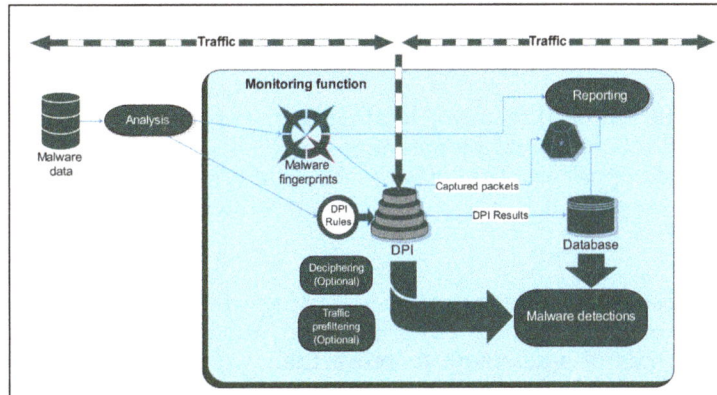

The layout of the experimental testbed for deep packet inspection (DPI) component.

In the evaluation environment, about 5 percent of all data flows were interpreted as HTTP application flows by the DPI engine and therefore were compared with fingerprints (i.e., signature detection). In real time analysis we could not measure any increase of CPU usage compared to the reference DPI analysis when the same number of metadata attributes were extracted. It has been found that the actual performance penalty should be measured at high data rates when packet drops may occur due to the additional processing.

The average processing delay was defined as the measure of the delay from the time the flow starts to the time a packet is received that allows the first detection decision to be made. In the test bed environment, the average detection decision delay was 57 ms. The results of the validation show that it is possible to perform DPI in SDN scenarios by using the proposed DPI component.

INDEX

A

Access Point Name, 45, 143

Authentication Center, 34, 87, 136, 143

B

Bandwidth, 6, 36, 41-42, 48-49, 56, 59, 61-63, 70, 73, 75, 86, 89-90, 95-97, 108-109, 112, 149, 172-181, 183, 191-192, 194

Base Station Controller, 34, 87, 102, 110, 113, 123-124, 129, 131, 137, 140

Base Transceiver Station, 46, 87, 100, 112-113, 122, 131, 140, 173

Battlefield Networks, 10

Bit Error Rate, 41, 43

Blocking Probability, 8-9, 23

C

Call Admission Control, 1, 8, 149, 152

Call Bounding, 8, 10, 13

Call Thinning Schemes, 9

Carrier Aggregation, 57, 60-61

Cell Site, 1, 69, 100, 102

Cell Tower, 5-6, 68, 100-103, 116, 121, 126

Cellular Network, 1, 7-9, 17, 64, 67, 88, 101, 107, 111, 115, 127, 132, 140

Cellular Router, 85

Circuit Switched Data, 43, 47, 74, 133

Code Division Multiple Access, 6, 49, 67, 70, 73, 75, 77, 79, 183

Cognitive Radio, 51-52, 61, 97

D

Data Transfer Rates, 69, 74

Deep Packet Inspection, 4, 200, 206, 217

Digital Encryption, 67-68

Digital Modulation, 37, 49

Digital Signal Processing, 20, 53

Doppler Shift, 41

Dropping Probability, 8-9

Dual Transfer Mode, 45

E

Erlang Formula, 21, 30

F

Femtocells, 61, 95, 103-110, 166, 170, 183

Fractional Dynamic Reservation, 10-11, 14

Frequency Band, 34, 49, 57, 79, 96-97, 160

Frequency Division, 34, 37, 46, 49, 54, 81

Frequency Division Multiple Access, 34, 49

G

General Packet Radio Service, 43, 135, 144, 184

Global Positioning System, 36

Guaranteed Bit Rate, 63, 143, 148

H

Handover Calls, 8-10, 16-17

High Chip Rate, 79, 86

Home Location Register, 34, 87, 130, 135, 138

I

Internet Of Things, 97, 190

Internet Protocol, 44, 62, 89, 141, 147, 175, 183

L

Long-term Evolution, 6, 51, 53, 55, 93, 97

Lossy Compression, 68

M

Media Gateway, 87, 134

Mesh Routing, 97

Mobile Modems, 69

Mobile Network, 1, 5-7, 41, 46, 67-68, 73, 75, 100, 104, 106, 110, 112, 124, 130, 135, 146, 164-165, 171-176, 183-189, 191, 193-199, 203, 207

Multi-tier Cellular Networks, 9, 16-17

Multi-user Detection, 43

Multimedia Messaging Service, 44-45, 139

N

Network Access Control, 3

Network Address Translation, 96, 184, 204

Network Coverage, 5, 50, 68, 101, 103, 168

Network Management System, 2, 113, 123

Network Time Protocol, 109

Nordic Mobile Telephone, 65-66

O

Open Systems Interconnection, 49
Open Wireless Architecture, 96
Orthogonal Codes, 37-38, 40-41
Overflow Streams, 25, 27

P

Packet Switching, 56-57, 96
Paging Channel, 35
Phase-shift Keying, 95
Picocell, 104, 110-111, 123
Poisson Traffic, 21, 24

Q

Quality Of Service, 10, 43, 51, 62, 108-109, 142, 148

R

Radio Access Network, 54, 63, 74, 82, 134, 146
Random Traffic Method, 10, 20
Real-time Communications, 98

S

Session Initiation Protocol, 63, 106
Short Message Service, 45, 63, 68, 131, 135
Single Tier Network, 14, 19
Spatial Division Multiple Access, 52, 94
Spread Spectrum, 35-36, 41, 70, 79-80, 89
Static Reservation, 11-16, 18
Statistical Multiplexing, 46, 50-51, 95

T

Time Division Multiple Access, 34, 43, 49, 67, 76, 79
Transmitter, 33, 35, 64, 77, 96, 102, 105, 131, 160, 167

V

Virtual Reality, 98

W

Wireless Application Protocol, 44-45
Wireless Communication, 1, 7, 9, 101, 113, 182

Z

Zero Generation, 64-65

www.ingramcontent.com/pod-product-compliance
Lightning Source LLC
Chambersburg PA
CBHW082045190326
41458CB00010B/3464